# An Annotated Calendar
# of the
# LETTERS OF CHARLES DARWIN
# in the Library of
# The American Philosophical Society

Figure 1. Charles Robert Darwin, aged 66. This proof of an etching was executed by Paul Adolphe Rajon (1842/43-1888) after a portrait from life painted in oils in 1875 by Walter William Ouless (1848-1933) for the Darwin family. A replica by the artist of the oil painting is at Christ's College, Cambridge. The marginal sketches are practice dry point engravings which constitute the remarque of the proof.

Charles Darwin said of this likeness: "I look a very venerable, acute, melancholy old dog; whether I really look so I do not know." His son Francis called it "the finest representation of my father that has been produced." (American Philosophical Society)

# An Annotated Calendar of the LETTERS OF CHARLES DARWIN in the Library of The American Philosophical Society

Edited and Annotated
by
P. Thomas Carroll

Foreword
by
Frederick Burkhardt

SR Scholarly Resources Inc.
1508 Pennsylvania Avenue · Wilmington, Delaware 19806

First published 1976
Printed in the United States of America

Scholarly Resources, Inc.
1508 Pennsylvania Avenue
Wilmington, Delaware 19806

Library of Congress Catalog Card Number: 75-29739
International Standard Book Number: 0-8420-2077-2

Inquiries concerning the content of this book may be sent to:
The American Philosophical Society
The Library
105 South Fifth Street
Philadelphia, Pennsylvania 19106

*In memory of*
*my father*
*who taught me the beauty*
*and the intricacies*
*of the*
*English language*

# Contents

# Illustrations

# Foreword

The Library of the American Philosophical Society took its first step toward assembling a collection of manuscripts and books relating to Charles Darwin and evolution in 1950. In that year, at the suggestion of William E. Lingelbach and with the support of Edwin G. Conklin, librarian and president respectively, the Society purchased at auction in London 177 letters from Darwin to Sir Charles Lyell and some 277 more to Lyell from other British and European scientists. Copies of the letters were deposited in the British Museum, and the originals became the nucleus of the Society's Darwin collection. (See Edwin G. Conklin, "Letters of Charles Darwin and Other Scientists and Philosophers to Sir Charles Lyell, Bart.," *Proc. Am. phil. Soc., 95* [1951], 220-22.)

To extend the collection Dr. Lingelbach asked the assistance of Dr. Loren C. Eiseley, who was then writing *Darwin's Century.* For some years Dr. Eiseley systematically scanned catalogues, made lists of desiderata, and, on visits to booksellers here and in England, purchased books by and about Darwin. Meanwhile the Library continued to acquire letters of Charles Darwin and other 19th century naturalists until it now has almost 700 by Darwin alone.

In 1973, the preparation of a calendar was begun by the Library in the belief that scholars could use a guide that was something more than a list but short of fully edited transcriptions of the Darwin letters. The work has been supported by a grant from the Andrew W. Mellon Foundation and it is now published with the approval of Mr. George P. Darwin.

The calendar was prepared by P. Thomas Carroll. Besides providing concise descriptions of the contents of unpublished letters, Mr. Carroll has indicated when changes were made in published versions, corrected erroneous transcriptions and added notes. A particularly important contribution of this calendar is the method Mr. Carroll and a colleague, Professor Thaddeus Trenn, have devised for dating the letters more accurately. All Darwin scholars are familiar with this difficult problem and all of them will appreciate help with it.

The present volume contains a substantial portion (about 15%) of all the Darwin letters that have so far been located. A search is now in progress to find as many more as possible for a "Collected Letters of Charles Darwin" of which I am a Co-Editor with Dr. Sydney Smith of St. Catharine's College, Cambridge. On the basis of my experience, I can attest to the skill and thoroughness with which Mr. Carroll has dealt with the many difficulties that confront an editor of Darwin's correspondence.

Scholarly interest in Darwin has been growing steadily in recent years and the Library of the American Philosophical Society has become a center of Darwin research on this side of the Atlantic. The present volume is thus a welcome introduction to a collection that has become an indispensable resource for research on the life and work of Charles Darwin and the history of the theory of evolution.

Frederick Burkhardt

# Acknowledgments

Despite indications to the contrary on the title page, this calendar was by no means the work of one or even two persons; a great many acknowledgements—too many, in fact, for all of them to be explicit here, and too deeply felt by the editor for adequate expression in words—are in order.

Above all else, of course, this calendar would not have been possible had it not been for the prodigious efforts of Charles Darwin himself, and so this book is his. By poring over his letters, I believe, I have come to know him fairly intimately, and I can assure my readers that, in this case at least, the old adage about familiarity breeding contempt is inapplicable. I know that this admission bodes ill for my retention of a critical historical perspective, but it would be a worse sin for me to feign objectivity. Besides, if my readers study these letters as I have, I think they will agree in all fairness that we should all be grateful that such a wonderful man as Darwin was once among us, however briefly and reclusively.

I am equally grateful to Darwin's great-grandson, Mr. George P. Darwin, for permission to produce this calendar and to quote extensively from the letters, regardless of the intimacy of their contents; this graciousness demonstrates that devotion to rigorous scholarship is as much the Darwin hallmark today as it was a century ago.

Financial support for the preparation of this calendar was provided by a generous gift to the American Philosophical Society by the Andrew W. Mellon Foundation.

The commodious facilities of the Library of the American Philosophical Society are a scholar's delight; I am grateful to the Society for the privilege of using them. Each of my colleagues on the Library staff has given generously of his or her time and expertise to the production of this work. Those who have contributed directly to the final manuscript were Helen Black, who conscientiously typed both my handwritten transcriptions of the letters and the manuscript of the front matter, and B. Dodelin, who prepared the photographic prints for the illustrations. Everyone else, each in his or her own way, not only added to this work but also made my stay at the Library pleasurable. I regret that there is not enough space to name them individually.

Many other individuals aided this project. I am grateful to Dr. Frederick Burkhardt, President Emeritus of the American Council of Learned Societies, for writing the foreword. Special thanks go to Dr. Sydney Smith, Lecturer in Zoology, St. Catharine's College, Cambridge University, and to Mr. Peter Gautrey, Cambridge University Library; with astonishing skill and resourcefulness they helped to date and otherwise identify the most intransigent letters in the collection. Similarly, Dr. Thaddeus J. Trenn, Department of the History of Science, University of Regensburg, helped considerably in the identification of letters from Darwin's cirripede period; in addition, as I discuss in the Introduction, Dr. Trenn has been my collaborator in the attempt to make sense of the many variant Down House addresses in the headings of Darwin's letters. Frederick Burkhardt has helped with the identification and arrangement of some of the letters. Professor Malcolm J. Kottler, Department of Ecology and Behavioral Biology, University of Minnesota, shared with me all of his determinations of the dates and the instances of publication of the Darwin-Romanes letters. Dr. Whitfield J. Bell, Jr., Librarian of the American Philosophical Society, not only supervised the project but also arranged for the publication of the work. In addition to the aid provided to me as part of their regular duties, Mr. Carl F. Miller, Assistant Manuscripts Librarian, and Mr. Murphy D. Smith, Associate Librarian, American Philosophical Society Library, were especially helpful in deciphering the more exotic examples of Darwin's difficult handwriting. The latter also offered many useful suggestions concerning format of entries. Whitfield J. Bell, Jr., Jeffrey L. Sturchio, and Frederick Burkhardt provided criticisms of early drafts of the front matter; Frederick Burkhardt also helped to proofread the calendar entries. Mr. Michael Glazier, President of Scholarly Resources, Inc., has been a generous and patient publisher. Robert F. Bud, Department of the History and Sociology of Science, University of Pennsylvania, helped to investigate British copyrights of the illustrations. Professor Leonard G. Wilson, Department of the History of Medicine, University of Minnesota, helped to publicize the calendar among Darwin scholars.

The following individuals aided in transcribing, dating, locating, and/or annotating one or more letters: Marianne Abel, Academy of Natural Sciences of Philadelphia; John Angell, Free Library of Philadelphia; Sir Hedley Atkins, M.D., Down House; Suzanne W. Brown, Chicago Academy of Sciences; Colin Burton, Baird & Tatlock (London) Ltd.; Dr. Ralph Colp, Jr.; Professor Joseph Ewan, Department of Biology, Tulane University; Ellen G. Gartrell, Assistant Curator, Historical Collections, College of Physicians of Philadelphia; Sheila K. Hart, Harvard College Library; F. A. Milligan, Hereford and Worcester County [England] Libraries; Professor James A. Rogers, Department of History, Claremont Men's College; Jeffrey L. Sturchio, Department of the History and Sociology of Science, University of Pennsylvania; Margaret Teransky, Free Library of Philadelphia; Philip Titheradge, Down House; and Professor David B. Wilson, Department of History, University of Oklahoma. My colleague and friend Roy Goodman here in the Library is unparalleled as a reference librarian.

In the course of my research, I have enjoyed the facilities of and/or received the competent assistance of the staffs of the Academy of Natural Sciences of Philadelphia, the Burndy Library, the Chicago Academy of Sciences, the Harvey S. Firestone Library of Princeton University, the Free Library of Philadelphia, the Harvard College Library, the Hereford and Worcester County [England] Libraries, the Historical Society of Pennsylvania, and the Charles Patterson Van Pelt Library of the University of Pennsylvania. I am also grateful to all those persons and institutions mentioned in the second appendix for making its compilation possible. A copy of the Darwin letter to Gray which I quote in the introduction was provided by the Gray Herbarium, Harvard University, which has the original. Those who have graciously given permission for reproduction of one or another of the illustrations are acknowledged where the figures appear.

Preparation of this calendar has consumed much of my time and attention over the last two years. Accordingly, I should like to thank the faculty, the staff, and my fellow graduate students in the Department of the History and Sociology of Science at the University of Pennsylvania for tolerating my reduced visibility around the Department. Similarly, my wife Nan, my late father, and my mother have suffered me cheerfully whenever the excesses of my enthusiasm for Darwin and the history of science have overflowed into my family life.

Many others, too numerous to mention, have aided the compilation of this volume in some way. Despite all this help, endless factual and interpretive errors, of which I am painfully aware and for which I am solely responsible, undoubtedly remain. I hope that scholars will both forgive me for these and report them to the Library for the benefit of future students of Darwin.

P. T. C.

# Introduction

Thousands of pages have been written about Charles Robert Darwin (1809-1882), whose *Origin of Species* laid the foundation for modern evolutionary theory. Even so, our picture of Darwin and our assessment of his place in history are undergoing constant revision. Relatively speaking, we are, after all, only recently accustomed to the notion that we share a common ancestry with apes, and few outside the biological professions yet appreciate the sweeping implications of population thinking (as opposed to typological thinking) or the concept of ecology. The social and political significance of Darwin's work will probably be debated for some time to come. Many historians of science are interested in the processes by which Darwin's ideas were formulated, disseminated, and ultimately accepted by scientists and by laymen. Scientists continue to be enlightened by reading Darwin,[1] and publication of transcriptions of Darwin's various manuscripts occurs regularly.[2]

Given these developments, as well as the added stimulus provided by the 1959 centennial of the publication of the *Origin,* it seems certain that research and writing upon Darwin will continue, with scholars demanding access to ever more records of Darwin's life and work. This calendar was prepared with this demand in mind, so that readers may better interpret the significance of this great naturalist's achievements and those of his colleagues.

The five remaining sections of this introduction explain in a general way the design and the intended utility of the calendar. They discuss the following topics: 1)the primary importance of letters in the assessment of Darwin's life and work; 2) the justification for publishing these letters; 3) the nature of the Darwin letters collection at the Library of the American Philosophical Society; 4) the reasons for publishing the letters in the more-or-less unfashionable calendar format; and 5) the unique problems encountered in dating the letters. A rather extensive and detailed statement of the editorial method employed in this calendar follows the introduction.

### *Darwin's Letters*

Modern man, armed with the telephone and often inundated with bureaucratic paperwork, might find it hard to comprehend how important letters were to life among the literate elite of the nineteenth century, and how valuable these nineteenth-century letters now are to scholars. But Darwin's contemporaries knew well the value of letters; Thomas Jefferson put the point succinctly in 1823 when he argued that "the letters of a person, especially of one whose business has been chiefly transacted by letters, form the only full and genuine journal of his life."[3] This

1. This is no place for a systematic review of the literature, but here are some examples: on the origins of Darwin's ideas, see recent articles by Barbara G. Beddall, Sandra Herbert, and Joel S. Schwartz in the *Journal of the History of Biology,* and Camille Limoges, *La sélection naturelle: Étude sur la première constitution d'un concept (1837-1859)* (Paris: Presses Universitaires de France, 1970); on Darwin's method, see Michael T. Ghiselin, *The Triumph of the Darwinian Method* (Berkeley: Univ. of California Press, 1969), and Stephen Jay Gould, "Darwin's 'Big Book'," *Science, N.Y., 188* (1975): 824-26; on the dissemination and early reception of Darwin's ideas, see Thomas F. Glick, ed., *The Comparative Reception of Darwinism* (Austin, Texas: Univ. of Texas Press, 1974), and David L. Hull, *Darwin and His Critics: The Reception of Darwin's Theory of Evolution by the Scientific Community* (Cambridge, Mass.: Harvard Univ. Press, 1973); on the ultimate acceptance of Darwinism, see William B. Provine, *The Origins of Theoretical Population Genetics* (Chicago: Univ. of Chicago Press, 1971); on the enlightenment of modern scientists, besides the numerous examples in population genetics, see Paul Ekman, ed., *Darwin and Facial Expression: A Century of Research in Review* (New York: Academic Press, 1973).

2. Some recent published transcriptions of Darwin's manuscripts include, in chronological order of publication: Barlow, ed., *Autobiography; Darwin and Henslow*; Howard E. Gruber, *Darwin on Man: A Psychological Study of Scientific Creativity, Together with Darwin's Early and Unpublished Notebooks,* transcribed and annotated by Paul H. Barrett (New York: E. P. Dutton & Co., Inc., 1974); and Stauffer, ed., *CD's Nat. Selection.* This list is hardly exhaustive. See note 16 for recent printings of Darwin's letters.

3. Thomas Jefferson to Robert Walsh, April 5, 1823, as quoted in *The Papers of Thomas Jefferson,* ed. Julian P. Boyd, *et. al.* (Princeton, N.J.: Princeton Univ. Press, 1950-), I, xi.

was especially true of scientists and other scholars of the day, for whom communication of the written word was such an integral part of intellectual activity.

Letters are particularly significant in Darwin's case. Physically isolated at Down House for most of his life and often immobilized by chronic illness, Darwin depended more heavily than others upon letters as his principal mode of communication. Moreover, because correspondence was Darwin's usual medium of contact with his closest friends, his letters often exhibit a degree of candor about his life and work not revealed clearly in any of his other papers and publications—not even in the intimate *Autobiography* which, though written for his own satisfaction and the entertainment of his children, lacked the presence of daily communication and was tempered somewhat by the restraints of Darwin's time and class.[4] Finally, the unusually heterogeneous quality of the Darwin letters reveal better than any other of the records of his life the diversity of his daily routine and the catholicity of his interests and preoccupations.

Scholars who examine the record of Darwin's life contained in his letters can expect to find many types of information. For the historian of science with a bent toward the history of ideas, there is abundant material on Darwin's scientific views. Sometimes a passage in a letter clarifies or corrects a claim in one of Darwin's published works; for example, in a letter to Charles Lyell in 1860, Darwin is tempering a rather extreme contention, which he had written into the *Origin,* when he writes that "Ammonites have become wholly extinct in a remarkably sudden manner *relatively* to most other families [and not absolutely, as implied in the *Origin*]; I meant only this [in the *Origin*], but I see I have not been nearly guarded enough."[5] At other times Darwin retracts assertions made in previous letters; in some cases only the earlier, unamended claim has been published, so that the unsuspecting scholar may be misled if he does not peruse carefully the unpublished materials.[6] Occasionally, a Darwin letter will contain opinions and theories not yet expressed in print, or at least not yet expressed as well as in the particular unpublished letter at hand; writing to Lyell in 1860, for example, Darwin argued that "how far to lump & split species is indeed a hopeless problem. It must in the end, I think, be determined by mere convenience."[7] It would be hard to locate a more succinct or more modern expression of Darwin's well-known disbelief in the existence of identifiable species.

Scholars interested in the social history of science, the growth of scientific ideas, the sociology of science, and the dynamics of scientific communities also will find much of interest in the letters, for they contain many of Darwin's views on these subjects, especially in reference to Darwin's assessment of the prospects for ultimate acceptance of his own theories. This is best exemplified in Darwin's confidences to friends in 1859 and 1860 regarding the fate of the *Origin,*[8] but there are other examples concerning Darwin's views about priority and his

---

4. Nora Barlow, ed., *Autobiography.*

5. Darwin to Lyell, Sept. 12, [1860], APS; see below, p. 85. (This and other examples of Darwin's letters used in the introduction are selected from *unpublished* portions of letters now in the American Philosophical Society; they illustrate the wealth of material edited out of previously published versions of Darwin's letters.) The passage on Ammonites which is being corrected is in Darwin, *Origin* (1859 *or* 1860), 321-22; according to Peckham, *Variorum Origin,* 531-32, lines 78-83, Darwin never brought his published claim regarding Ammonites into agreement with the admissions made to Lyell in this letter.

6. For example, in a letter to Lyell dated November 18, 1849, and printed in *More Letters,* II, 130-31, letter 486, Darwin wrote that "without most distinct evidence I will never admit that a *dike* joins on rectangularly to a stream of lava." Sixteen days later, in a letter dated December 4, he retreated somewhat from this position: "I remember in my last letter talking very big about dikes never being connected directly (i.e. rectangularly) with lava-streams; but it is clear that such occur frequently at the Sandwich Is[lan]ds without any cones." Although he continues by claiming that the situation at the Sandwich Islands [i.e. Hawaii] is "a rare exceptional case", Darwin clearly is amending his earlier claim. The December 4 letter has not been published before. See below, pp. 32-34.

7. Darwin to Lyell, December 4, [1860], APS, see below, pp. 89-90.

8. Just after the *Origin* was published, during the period when reviews of the book began to appear, Darwin told Asa Gray, "I have made up my mind to be well abused; but I think it of importance that my notions sh[oul]d be read by intelligent men, accustomed to scientific argument though *not* naturalists. It may seem absurd, but I think such men will drag after them those

estimation of the role his cirripede work would play in the growth of science.[9] Still another revealing type of information expressed only in the letters is that relating to Darwin's assessments of his colleagues—enlightening glimpses of Darwin's critical insight at work which rarely surfaced in the polite society of contemporary England but are now absolutely necessary for a full understanding of his professional identity and behavior. Where else can one find Darwin saying of Robert FitzRoy: "I never cease wondering at his character, . . .full of good . . . traits but spoiled by such an unlucky temper. Some part of . . . his brain wants mending. . . .?"[10] It is no wonder that this passage, as well as one which describes Richard Owen as "wonderfully clever in his malevolence,"[11] usually were deleted silently from the earlier published versions of Darwin's letters—this is all the more reason for including them here.

To complement such insights and revelations, there is much autobiographical matter in the letters. It might be used to construct a reasonably good record both of Darwin's face-to-face encounters with his colleagues and of the content of such meetings.[12] Often the letters record Darwin's whereabouts, which information is especially useful in determining the dates of Darwin's frequent one-day trips into London—something not systematically recorded elsewhere.[13] Moreover, the letters constantly remind us that Darwin was as much a family man and a medical patient as he was a naturalist and a scientist, and they show how these different aspects of Darwin's life tended to interact. For example, in 1860, Darwin's daughter

---

naturalists, who have too firmly fixed in their heads that a species is an entity." (Darwin to Asa Gray, December 21, [1859], Gray Herbarium, Harvard University; printed in *Life and Letters,* II, 244-45; text above from manuscript letter.) In this passage, Darwin implied that his evolutionary "notions" were revolutionary in nature, at least among naturalists, and that he expected his supporters to come from *outside* the community of naturalists. How important a role he foresaw for these supporters is revealed four months later in his remark to Lyell that "the non-comittal [*sic*] men do not always most help a science." (Darwin to Lyell, April 27/28, [1860], APS, see below, p. 76.) Darwin reiterates this view and identifies some of his supporters in May when he writes to Lyell: "I can very plainly see, as I lately told [Joseph Dalton] Hooker, that my Book would have been & [would] be a mere flash in the pan, were it not for you, Hooker & a few others." (Darwin to Lyell, May 18, [1860], APS, see below, p. 78.) A month later—ironically on the eve of the historic Oxford meeting of the B.A.A.S. at which the *Origin* was debated so hotly—Darwin despaired of the rapid conversion of naturalists through the intercession of supporters, arguing instead that "time alone will bring naturalists round, when they find that they can explain many facts on such views as mine, & cannot on view of creation." (Darwin to Lyell, [June] 25, [1860], APS, see below, p. 82.) It is almost as if Darwin's practice of population thinking, so well-used on plant and animal species, was being applied to scientific communities; social historians and sociologists of science may find this of some interest, as might some philosophers of science.

The foregoing is not meant, of course, to delineate the recent historiography of the social side of Darwin studies; for a stimulating marxist treatment of some aspects of this, see Robert Young, "The Historiographic and Ideological Contexts of the Nineteenth-Century Debate on Man's Place in Nature," in Mikuláš Teich and Robert Young, eds., *Changing Perspectives in the History of Science: Essays in Honour of Joseph Needham* (London: Heinemann, 1973), 344-438, esp. 361-88.

9. In a confidence to Lyell in 1860, Darwin revealed his views on the role of priority in science, exposing in the process why he apparently never felt completely comfortable publishing simultaneously with Alfred Russel Wallace in 1858. Regarding a reference to Darwin by Asa Gray in one of the latter's works, Darwin wrote: "he put my name before [that of Edward] Forbes on Glacial distribution: & I told him in answer that I had written out the notion 3 or 4 years before Forbes, but that I had no sort of claim to notice on this head, as he published first, & that in the Origin I shd. of course take no notice of this." (Darwin to Lyell, [February] 12, [1860], APS, see below, p. 71.) This gives some indication of Darwin's scrupulous professional ethics.

On cirripedes, Darwin's expectations regarding the support he would receive for his radical discoveries were far more limited than were his expectations for his work on the origin of species. When Albany Hancock informed Darwin that he believed in Darwin's discovery of complemental males, Darwin replied that he had "greatly feared tha no one would believe in them; and now I know that [Richard] Owen, [James Dwight] Dana, and yourself are believers, I am most heartily content." (Darwin to Hancock, January 10, [1853], as printed in John Hancock, [ed.], "Letters from C. Darwin, Esq., to A. Hancock, Esq.," *Nat. Hist. Trans. Northumb., 8,* pt. 2 [1886]: 250-78, at 269; original at Historical Society of Pennsylvania, Philadelphia.)

10. Darwin to Lyell, August 9, [1838], APS, see below, p. 4. Although Darwin probably was not anticipating it here, FitzRoy committed suicide in 1865.

11. Darwin to Lyell, July 30, [1860], APS, see below, p. 83.)

12. In 1860, for example, he wrote Lyell: "I have nothing to say, as I have seen no one (except indeed [Joseph Dalton] Hooker for an hour or two at Kew) for an age." (Darwin to Lyell, July 30, [1860], APS, see below, p. 83.) This letter implies that the content of Darwin's meetings with colleagues forms much of the meat of his letters. Still, Darwin exaggerates in his claim that, when he has seen no one, he has nothing to say.

13. For example, we can tell from the letters that Darwin met Lyell in London on February 6, 1845. (Darwin to Lyell, Saturday, [February 8, 1845], APS, see below, pp. 16-17.) Unfortunately, this type of information is sometimes incorrect, either because Darwin would announce a date for a trip to London and then be too ill to travel on the appointed day, which occurred regularly, or because Darwin got his dates mixed up, which occurred very occasionally.

contracted remittent fever, and the family was forced by their concern for the young girl's life to remove to a healthier climate than that found at Down. While thus away, Darwin found himself unable to continue the research he had been conducting at home, so he filled his idle time by examining the sun-dew, an insectivorous plant common to the area being visited. This work eventually led to the publication of an entire book on insectivorous plants, and one could therefore argue the extreme externalist position that illness had been the cause for its appearance.[14]

Still another type of correspondence consists of purely business letters, which range from orders and acknowledgments of books and reprints—data valuable to scholars—to routine details such as the addition of a wing to Down House in 1876-1877 or the purchase of medical supplies.[15] Such routine letters show, if nothing else, Darwin's meticulousness, and they are sometimes useful for dating other Darwin letters of greater import.

*Need to Publish Letters*

To a certain extent, scholars have been reaping benefits from Darwin's letters for long time, since the bulk of Darwin's most important letters have been in print for well over fifty years, and additional letters have been published from time to time ever since.[16] One might well ask, then, why a calendar of 700 more letters is necessary, especially since portions, at least, of roughly half of the letters in the American Philosophical Society's collection have already been printed. Several different responses can be given to this question.

Most importantly, there is the problem of selection criteria in the major published collections of Darwin's correspondence. The three works which contain published versions of many of the items calendared below (as well as a great many other Darwin letters) provide only a selection of Darwin's correspondence, and those letters which are included are usually not transcribed in full. In all three of these works, the choice of both which letters are to be included and which portions of the chosen letters are to be printed has been colored by the motives of the editors. As these motives are not those of historians today, there are limitations—some of them severe—to the usefulness of these works as definitive references for modern scholars. To illustrate the point: "In choosing letters for publication," says Darwin's son Francis in the opening line of the preface of his edition of his father's *Life and Letters,* "I have been largely guided by the wish to illustrate my father's personal character." In some cases, such guidance led him to give a relatively low priority to the illustration of Darwin's scientific work, and in most cases, it meant deleting family matters as well as derogatory remarks and

---

14. On his daughter's illness, see letters to Lyell during summer of 1860 (Darwin to Lyell, May 18, [1860], *et seq.*, APS; see below, p. 78 ff). See especially: Darwin to Lyell, July 30, [1860], APS, see below, p. 83; and *ibid.*, August 11, [1860], APS, see below, pp. 83-84. The first chapter of Darwin's book on insectivorous plants begins: "During the summer of 1860, I was surprised by finding how large a number of insects were caught by the leaves of the common sun-dew (*Drosera rotundifolia*) on a heath in Sussex. I had heard that insects were thus caught, but knew nothing further on the subject." (Darwin, *Insectivorous Plants* [1875], 1.)

15. On the addition to Down House, see: Darwin to [William] Marshall, September 19, 1876, APS, see below, p. 173, *ibid.*, September 29, [1876], APS, see below, pp. 173-74; and *ibid.*, November 22, [1876], APS, see below, p. 174. On the purchase of an enema, see Darwin to ?, November 8, [1871-1875], APS, see below, p. 149.

16. A great many Darwin letters appear in the three volumes of *Life and Letters,* the two volumes of *More Letters,* and the two volumes of *Emma Darwin.* A few of the many locations in which Darwin correspondence has appeared recently are, in chronological order: Gavin de Beer, "Further Unpublished Letters of Charles Darwin," *Ann. Sci., 14* (1958): 83-115; *idem,* ed., "Some Unpublished Letters of Charles Darwin," *Notes Rec. R. Soc. Lond., 14* (1959): 12-66; Robert M. Stecher, "The Darwin-Innes Letters: The Correspondence of an Evolutionist with His Vicar, 1848-1884," *Ann. Sci., 17* (1961): 201-58; *Darwin and Henslow*; Gavin de Beer, ed., "The Darwin Letters at Shrewsbury School," *Notes Rec. R. Soc. Lond., 23* (1968): 68-85; Robert M. Stecher, "The Darwin-Bates Letters: Correspondence between Two Nineteenth-Century Travellers and Naturalists," *Ann. Sci., 25* (1969): 1-47, 95-125; Paul H. Barrett and Alain F. Corcos, "A Letter from Alexander Humboldt to Charles Darwin," *J. Hist. Med., 27* (1972): 159-72; Barbara G. Beddall, " 'Notes for Mr. Darwin': Letters to Charles Darwin from Edward Blyth at Calcutta: A Study in the Process of Discovery," *Journal of the History of Biology, 6* (1973): 69-95; Thaddeus J. Trenn, "Charles Darwin, Fossil Cirripedes, and Robert Fitch: Presenting Sixteen Hitherto Unpublished Darwin Letters of 1849 to 1851," *Proc. Am. phil. Soc., 118* (1974): 471-91; and Lewis S. Feuer, "Is the 'Darwin-Marx Correspondence' Authentic?" *Ann. Sci., 32* (1975): 1-12. For a reasonably complete, but by no means exhaustive, bibliography of Darwin letters published somewhat earlier, see Gavin de Beer, ed., "Some Unpublished Letters of Charles Darwin," *op. cit.,* this note, 60-62 and 66.

other embarrassing details.[17] *More Letters,* edited by Francis Darwin and A. C. Seward, somewhat redresses the balance; its editors say that they prepared the volumes "to give as full an idea as possible of the course of Mr. Darwin's work," and they state further that their compendium contains "practically all the matter that it now seems desirable to publish. But," they continue, "at some future time others may find interesting data in what remains unprinted; this is certainly true of a short series of letters dealing with the Cirripedes, which are omitted solely for want of space."[18] Similarly, in the preface to her two-volume biography of her mother, which contains 66 of her father's letters, Darwin's daughter, H. E. Litchfield, confesses that she began to prepare a record of her mother's life because she thought it would be "of value to her [Emma Darwin's] grandchildren," and then she altered and enlarged the scope of the book as she became interested in "the personalities of the writers" of the letters printed therein.[19] While such criteria of inclusion and exclusion were perfectly respectable and common in their day, and scholars have profited greatly from all these works despite their limitations, there is still a need to treat Darwin's letters more systematically, particularly with the needs of historians of science and other scholars in mind.

A still more serious indictment of the principal sources of Darwin's correspondence is that those letters which do appear in them were not edited according to rigorous modern standards. "In printing the letters," Francis Darwin confesses about the *Life and Letters,* "I have followed (except in a few cases) the usual plan of indicating the existence of omissions or insertions." One wonders what the "few cases" were, and exactly how few they were. "I have not followed the originals," he continues, "as regards the spelling of names, the use of capital letters, or in the manner of punctuation."[20] Similarly, in *More Letters,* "dots indicate omissions, but many omissions are made without being so indicated."[21] In *Emma Darwin,* "many omissions are made without putting any sign that this has been done. Neither the punctuation nor the spelling has been rigidly followed. But the sense has never *wittingly* been altered, although occasionally a word *evidently* omitted has been added without putting any sign that this has been done."[22]

In addition to the problems relating to the imprecise editorial standards of these works and to standards of letter selection which stress Darwin's character to the neglect of his career and other aspects of his life, some scholars may have difficulty in using these words because: the letters are not placed in chronological order; in far too many cases are incorrectly dated or are not dated at all[23]; and the indexes are inadequate.

To a certain extent, these final criticisms can also be levelled at some of the small selections of Darwin letters published as journal articles with increasing frequency since the centennial of the publication of the *Origin* in 1859.[24] Moreover, the quality of the transcriptions in some of these articles is not very good.[25] While this calendar will not rectify all of these problems

17. *Life and Letters,* I, iii.

18. *More Letters,* I, viii.

19. *Emma Darwin,* I, vii. The count of 66 letters is from the privately printed edition, published in 1904; other editions are probably slightly different.

20. *Life and Letters,* I, iv.

21. *More Letters,* I, ix.

22. *Emma Darwin,* I, ix. Italics added.

23. See, for example, the statement of the method of dating used by Francis Darwin and Seward (*More Letters,* I, x). For an example of an improperly dated letter, see Darwin to Asa Gray, April 4, [1858]; this letter is dated "1859" in the first edition of *Life and Letters* ([1887], II, 154-55), but it is deleted from later editions, probably because Francis Darwin realized that it was improperly dated.

24. See note 16, above.

25. The usual reason for a poor transcription is the inability of the transcriber to read Darwin's handwriting, but still another reason is that thorough editorial standards are usually not employed for the transcription of the small number of letters usually included in these articles. A simple but important example is the use of parentheses instead of brackets around the

(most notably because it contains a rather small and arbitrary selection of those letters published in these articles), it should clarify all such problems for the letters with which it deals. Moreover, use of this calendar should at least streamline the kind of research which can be conducted only while one is visiting the Society's archives. This should save much research time.[26]

*Nature of the Society's Darwin Collections*

To a first approximation, Darwin's correspondents can be grouped according to their relationship to Darwin and their function as contributors to his work. A look at a few of these various types of correspondent may help readers to obtain a overview of the Darwin collections calendared here.

Among the correspondents, there are at least six types. 1. Members of the Darwin family. Letters to and from these persons are sometimes the most intimate of all of Darwin's letters, but they are not so numerous as letters to and from some correspondents in the other groups discussed below. This is probably because of Darwin's proximity to his family—except, of course, during his early life, particularly during the voyage of the *Beagle,* when letters home to his sister are some of the most revealing letters Darwin ever wrote. 2. Confidants, friends, and close colleagues. These persons were the figures most closely associated with Darwin's everyday work, and the letters to them are the most useful for historians. In some cases, because of Darwin's removal from London and need to keep in close touch with his most trusted colleagues, the collected letters to these figures form an almost daily record of Darwin's life and work. 3. Other colleagues. These were scientists, naturalists, and the like with whom Darwin had only professional relations. Although somewhat less cordial than letters to confidants, correspondence with other colleagues is similar in substance to that with confidants. 4. Informants. These were rather special colleagues who supplied the sedentary Darwin with the observations and other data needed to test various aspects of his scientific theories. Darwin was careful in his selection of informants; he always checked to be certain that each one was an expert in the subject about which he was inquiring. Darwin had such informants in all corners of the world, but he also called upon nearby friends, relatives, former schoolmates, and acquaintances for information on topics about which such persons had some knowledge. 5. Business associates. This category includes booksellers, publishers, chemical suppliers, solicitors, architects, and the like. Correspondence with them is sometimes useful as a record of Darwin's activities, especially as an indicator of Darwin's reading, and it usually shows Darwin's attention to detail. 6. Unsolicited correspondents. From such persons Darwin received many letters, notes, and specimens. These ranged from crank mail to information on scientific curiosities about which Darwin showed great interest.

All six of these groups are represented to a greater or lesser degree in the American Philosophical Society's collections, and since some of the finest examples are not easily identified by perusal of the calendar, given its chronological format, they are discussed below. Unfortunately, they give a necessarily one-sided view of the relationship between Darwin and his correspondents, because there are only eleven letters *to* Darwin in the Society's collections.[27]

---

editorially-added word "*Cryptophialus*" in the printed text of a letter to Albany Hancock, December 25, [1849], as printed in John Hancock, [ed.], "Letters from C. Darwin, Esq., to A. Hancock, Esq.," *Nat. Hist. Trans. Northumb., 8* (1886): 250-78, at p. 258; original at APS, see below, p. 34. The unsuspecting scholar without access to the orignal letter might conclude from the printed text that by 1849 Darwin had identified his Arthrobalanus specimen as a Cryptophialus; this Darwin had not done—and in fact probably did not do until 1853—and realization of this is central to an understanding of Darwin's cirripede work (Thaddeus J. Trenn, "Charles Darwin, Fossil Cirripedes, and Robert Fitch: Presenting Sixteen Hitherto Unpublished Darwin Letters of 1849 to 1851," *op. cit.,* note 16, *passim,* esp. 472-73 and 472n.

26. In some cases, a trip to Philadelphia might be avoided entirely by the acquisition of photocopies of letters of interest by mail—a practice which the Society wishes to encourage.

27. This unfortunate circumstance results from Darwin's habit of destroying letters received; this practice was not discontinued until 1862, and even after that date Darwin did not save all of his letters. Apparently he did not think of letters in the way Jefferson did when he wrote the passage quoted earlier. (*Life and Letters,* I, v.)

At the core of the Library's holdings are the letters to Charles Lyell, the father of uniformitarian geology and Darwin's "Lord Chancellor" for science.[28] Until his death in 1875, Lyell was one of Darwin's closest confidants. Many of Darwin's letters to him have been published, although most of the passages concerning geology, as well as a few other crucial sentences here and there, have been excised from the printed versions. Some of the unpublished letters are also of importance.

The calendar lists an unusually rich correspondence with two other confidants: George John Romanes, the physiologist, and John Maurice Herbert, one of Darwin's schoolmates at Cambridge. Begun in 1874, the Darwin-Romanes correspondence warmed over time, as Darwin came to trust and admire this young colleague. In a sense, Romanes filled a gap for Darwin which was created with the death of Lyell in 1875, although in the Darwin-Romanes relationship the roles of elder and follower were reversed for Darwin from what they had been with Lyell. The letters to Romanes are replete with details about Darwin's pangenesis hypothesis and the experiments designed by the two men to test it, as well as much material on animal intelligence, spiritualism, and other topics. The letters to Herbert begin as early as 1828 and provide important insight into Darwin's early life, including his voyage on the *Beagle.* While some of these letters have been published, a few significant portions of them have not, and readers with a particular interest in the views of the young Darwin will probably want to study the Herbert entries in the calendar with extra care.[29]

Informants are almost as well represented in the calendar as are confidants. The Library has a particularly good collection of letters to George Henry Kendrick Thwaites, the Ceylonese naturalist, concerning the flora and fauna of Ceylon and India as well as the expression of the emotions of savages in these regions. The collection illustrates well both the questions Darwin asked and the answers he expected to receive. The correspondence with Thomas Campbell Eyton, another of Darwin's schoolmates at Cambridge, is of similar interest, especially on topics relating to osteology and other morphological similarities, dispersal mechanisms, and domestic varieties. Both the Thwaites and the Eyton letters are indicative of the methods of gathering data used extensively by Darwin during and after the *Origin* period (ca. 1854-1875); the earlier part of these years, of course, represent Darwin's most important fact-gathering episode since his return from the *Beagle* voyage.

The collections of letters to the correspondents named above by no means comprise all of the Library's Darwin holdings, of course. There are a great many other letters, sometimes in groups to a single correspondent (e.g. to Philip Lutley Sclater or to John Phillips) and sometimes scattered through miscellaneous correspondence. Only careful reading of all the letters in a given period will give scholars a genuine feel for the totality of Darwin's life and work.

### *Choosing the Calendar Format*

In choosing the calendar format for this guide, the primary objective has been to bring scholar and documents as close together as possible, given the restraints of economy, of time, and of the unique character of the originals. Almost thirty years ago, the usefulness of a calendar for achieving this objective in situations such as exist with the Darwin letters at the Society was summed up well by archivist and local historian Morris L. Radoff of the Maryland Hall of Records:

> . . . the calendar is especially valuable in treating (1) material that is badly

28. Darwin meant that he respected Lyell's scientific and professional judgment above that of all others. See *Life and Letters,* II, 119.

29. See particularly Darwin to Herbert, June 2, 1833, APS, see below, p. 2. The passage in this letter which discusses carnations and peaches shows Darwin's exposure to such subjects at an early age, and indicates that Darwin probably also discussed such topics while at Cambridge.

and hopelessly disarranged, . . . (2) materials of great complexity, where the chronological arrangement demands a reasonable amount of preliminary study, . . . and (3) materials which must be brought together in the calendar from . . . several [collections] in the same depository. . . .[30]

The material in the Darwin collections in the Library fits all three of these desiderata. First, it is indeed rather disarranged, owing primarily to the piecemeal method of acquisition of most of it. Second, because Darwin usually failed to date his letters completely and sometimes failed to date them at all, chronological arrangement demanded more time for preliminary study than would normally be available to the average scholar interested in only a portion of the Society's Darwin holdings and unable to afford either a lengthy stay in the Philadelphia area or photostatic copies of considerable amounts of Darwin materials. Third, since about half of the letters are dispersed throughout the archives as separate collections of letters from Darwin to a single correspondent, scholars interested in an overview of Darwin's correspondence would have had to conduct a multitude of separate searches in the absence of a single master calendar.

This calendar is somewhat more extensive than most. A description of this more extensive type of reference work and a justification for its employment in this instance are summed up well in the *Harvard Guide to American History:*

> In this [more extended] form of calendar, a digest of each document is printed and the more important and interesting parts are quoted. Thus a large part of the expense of printing the documents in full is saved, yet the meat of them is rendered available to students. As Americans are such inveterate indexers, arrangers, and bibliographers, it seems strange that they have seldom used this cheap but excellent method of presenting the heart of a manuscript collection or archive to the public.[31]

Because of its extensive nature, the calendar should serve the needs of most scholars. In those cases in which it is not sufficient, however, the original manuscript letters are still on deposit at the Library, and photocopies and/or microfilms of the originals are available for a small fee on a special order basis. In addition, after a brief period of a few years during which they will be used for another project, a set of complete and annotated transcripts of the letters (i.e. transcripts prepared by the editor to facilitate the compilation of this calendar) will be available for study.

*Resolving the Dating Problem*

Anyone who has worked with Darwin's letters knows how difficult it often is to assign a correct and exact date to each of his letters. Francis Darwin and A. C. Seward stated the problem well when they wrote that "Mr. Darwin, who was careful in other things, generally omitted the date in familiar correspondence, and it is often only by treating a letter as a detective studies a crime that we can make sure of its date."[32] For a compiler of a calender of Darwin's letters, of course, the inconsistent dating creates the same problems and puzzles that Francis Darwin faced, and, like him, one has to employ a variety of methods of dating that would do honor to Sherlock Holmes. The usual techniques involving watermarks, postmarks, content analysis, and handwriting analysis were sufficient in most cases, but even after such methods had been employed the dates on a number of letters remained ambiguous.

---

30. "A Guide to Practical Calendaring," *American Archivist, 11* (1948): 123-40, at 127.

31. Frank Freidel, ed., *Harvard Guide to American History,* rev. ed., 2v. (Cambridge, Mass.: Belknap Press of Harvard Univ. Press, 1974), I, 23.

32. *More Letters,* I, x.

Fortunately, study of the variant addresses appearing in the headings of almost all Darwin letters written since 1842 has allowed the dating of most of the letters in the calendar, at least to within a few years.[33] The research on this subject, conducted in collaboration with Thaddeus J. Trenn, University of Regensburg, indicates that Darwin used at least nine variant letterhead addresses for his home in Down as tabulated in the table below and illustrated in Figure 2.

| Type | Address Variant | Dates Used |
|---|---|---|
| Type 1: | Down near Bromley Kent (written) | 1842 to 1845 |
| Type 2: | Down Bromley Kent (written) | 1843 to 1846<br>and 1855 to 1861 |
| Type 3: | Down Farnborough Kent (written) | 1846 to 1855 |
| Type 4: | Down.<br>Bromley. (printed)<br>Kent. S.E. | 1861 to 1869 |
| Type 5: | Down.<br>Beckenham ~~Bromley~~.<br>Kent. S.E.<br>(printed, with Bromley crossed out and Beckenham added in writing) | 1869 to 1871 |
| Type 6: | Down, (printed)<br>Beckenham, Kent. | 1871 to 1875 |
| Type 7: | Down Beckenham Kent (written) | 1874 (briefly) |
| Type 8: | DOWN,<br>BECKENHAM, KENT (printed)<br>RAILWAY STATION<br>ORPINGTON. S.E.R. | 1874 to 1881 |
| Type 9: | DOWN,<br>BECKENHAM, KENT (printed, with<br>(RAILWAY STATION parentheses added<br>ORPINGTON. S.E.R.) in writing) | 1878 to 1882 |

(The end points for these dates were determined by recording the addresses and the dates for all unambiguously-dated Darwin letters which could be found, both here and at Cambridge University Library—probably about 1,000 letters in all. A plausible explanation apparently can be advanced for every change of address, and in most cases, the switch from one address type to another is sudden. For the purposes of dating the letters in this calendar, however, the greatest possible latitude in the beginning and final dates of use of each variant has been assumed; this has produced the overlaps in the time periods given above. The research on the variant addresses continues, and the final results will probably be published elsewhere.)

One final note about dates is necessary. Often the date assigned to a particular letter in this calendar will differ by a few days from the date given in a published version of the same letter, especially those published versions appearing in *Life and Letters* or *More Letters*. This is

33. To the best of the editor's knowledge, this regularity in the Darwin addresses was first discussed and employed for dating purposes by Gavin de Beer; see de Beer, "Some Unpublished Letters of Charles Darwin," *op. cit.*, note 16, 13-14. As de Beer points out, many of the printed versions of the letters give the simple address of "Down", thereby obscuring the variant headings.

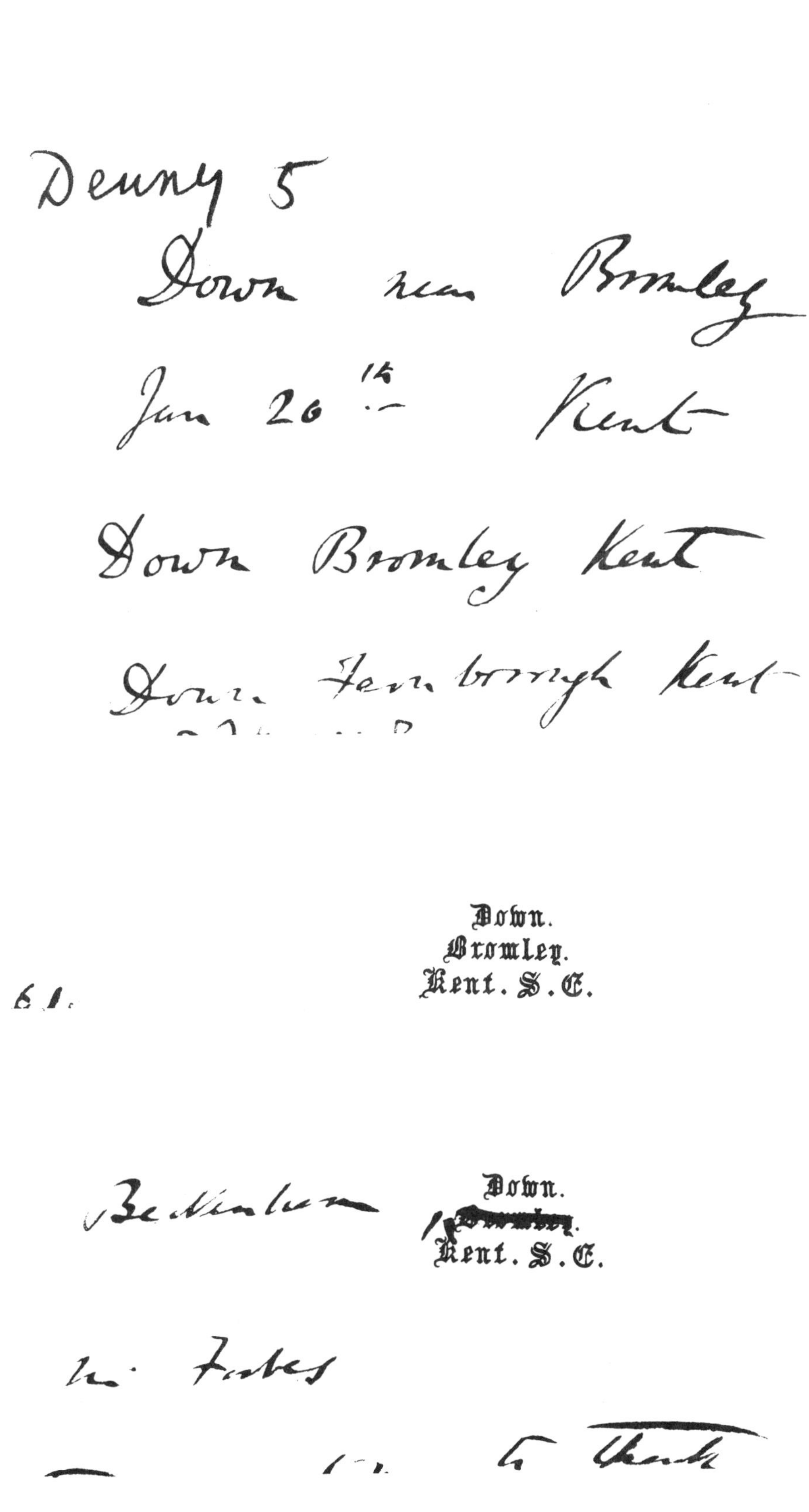


Figure 2. Variant Down addresses used by Darwin. (a) from top: Down near Bromley Kent (written); Down Bromley Kent (written); Down Farnborough Kent (written); Down. Bromley. Kent. S.E. (printed); the same, with Bromley changed to Beckenham by hand. (b) from top: Down, Beckenham, Kent. (printed); Down Beckenham Kent (written); Down, Beckenham, Kent. Railway Station Orpington. S.E.R. [for South Eastern Railway] (printed); the same, with the railway address in parentheses. The words "Denny 5" at the top of (a) were added in purple pencil by Francis Darwin as a means of identifying his father's letters during the preparation of the *Life and Letters.* See text for further details. (American Philosophical Society)

Down,
Beckenham, Kent.

Down Beckenham Kent

DOWN,
BECKENHAM, KENT.
RAILWAY STATION
ORPINGTON. S.E.R.

15 . 75

20 1881

DOWN,
BECKENHAM, KENT.
RAILWAY STATION
ORPINGTON. S.E.R.

ear Sir

eg leave to inform you that the
quired address, is
" A. R. Wallace Esq "

because the editors of these works often used the date of receipt of a letter, taken from the endorsement, rather than the actual date on which the letter was written.[34] In cases in which Darwin gave some reference to the day of the week on which the letter was written, it has been possible to determine the date of writing by reference to a perpetual calendar. In such cases, the correction of the date has been made silently.

P. T. C.

34. The editors state this explicitly in *More Letters,* I, x.

# Editorial Method

Purists might not find this calender to their liking. It deviates from traditional practices[35] in a number of details, from the distinctive overall format of each of its entries to an unusually liberal use of brackets and inverted commas. These innovations have been dictated by the uncommonly uniform nature of the items analyzed (almost all of them letters written by Darwin), by the peculiar characteristics of Darwin's letters, and by the unique needs of Darwin scholars. In the case of each uncommon procedure, a conscious determination was made that the benefits to be derived by departure from standard practice outweigh the perils and bother incumbent upon readers who must first school themselves in so many unorthodoxies.

Rules followed and usages observed are explained below, in five sections: format of entries–an extended guide to most of the practices followed; editorial conventions (i.e. treatment of orthography, punctuation, obsolete grammar, and the like in quoted passages); textual devices; descriptive abbreviations of items (e.g. ALS); and printing conventions. The first section, on format, explains what material is discussed in the four sections that follow it. Illustrations of sample manuscript letters are provided (Figs. 3 and 4), so that users may see for themselves how various situations were treated editorially.

*Format of Entries*

The calendar consists of a series of entries numbered consecutively in the left-hand margin. Except where noted, each entry represents one letter or other manuscript item, plus all accompanying enclosures, sketches, and the like. Unless indicated otherwise, all enclosures are in Darwin's hand, unsigned, and on approximately the same size and type of paper as the letters with which they are enclosed. Each entry consists of four parts: date; description of item(s); abstract, including publication information, if any; and annotations, if any. Each section is separated from the others by a blank line; otherwise, the entries are single space.

Arrangement of the parts of the entry in this way was decided upon to facilitate special treatment of certain unique aspects of the Darwin materials. Although the date of an entry is usually placed in the margin in other calendars, it has been given its own line in this calendar so that endorsements, postmarks, and watermarks could be included in full; scholars long accustomed to undated or ambiguously-dated Darwin letters will find this feature useful for checking the dates assigned to undated or incompletely dated letters. The part consisting of the date always contains the year, month, day, and (if provided by Darwin) the day of the week, in that order, on which the letter was written, followed by endorsements, postmarks, watermarks, and other information pertaining to date; such additional dating aids are always given as nearly exactly as possible to how they were on the original, and they are preceded by abbreviations identifying each (e.g. "wmk." precedes the watermark). Editorial conventions and textual devices used with the date are discussed below in two special sections on such matters.

The description, comprising the second part of each entry, is in two sections. The first section gives, as a first element, the full name of the other principal for the item besides Darwin, and it indicates whether this person is the author or the recipient of the item. This first element would thus read "To Charles LYELL" for a letter from Darwin to Lyell. Note

35. For the sake of argument, traditional forms are enumerated in: Morris L. Radoff, "A Guide to Practical Calendaring," *op. cit.*, note 30; and *idem*, "A Practical Guide to Calendaring," *American Archivist, 11* (1948): 203-22.

Some might also object to the method of production (i.e. photo offset of a carefully-proofed typescript, rather than typeset), but this was necessary to keep the cost per copy down to a level reasonable enough so that scholars could afford personal copies of the calendar.

that the last name of the principal is given entirely in upper case letters to facilitate scanning of the entries. (This is only done for individuals.) The second element of the first section of the description provides Darwin's address at the time of writing, as indicated by Darwin himself at the head of each letter. If the address is one of the variants of Darwin's Down address, it will be given as "Down", followed by an indication of the number of the variant type that appears (see p. xxiii, above). Thus, if Darwin wrote "Down Bromley Kent" on a given letter, for example, this will be indicated by "Down (type 2)". If Darwin's whereabouts are not indicated, this will be signified by "no location". If the stationery used has a black mourning border, or if an ink of a distinctive color was used by the author in the writing of the letter, this will be noted in parentheses following the indication of Darwin's location. The two elements of the first section are separated from each other by a semicolon.

The second section of the description begins a new line below the first section. It consists of three parts: the descriptive abbreviation for the item (e.g. ALS); the dimensions of the item, in inches, with vertical dimension given first[36] ; and the number of pages of text, rounded up to the nearest whole number of pages. Descriptive abbreviations used are discussed in a separate section below. If there is an envelope, enclosure, or sketch, or if there is an address for, or an endorsement by, the recipient, this will be indicated following the number of pages. If the text of the letter indicates that there should be an enclosure, but there is none, this will be indicated by the phrase "(enclosure wanting)" following the number of pages. If the address for or the endorsement by the recipient *occupies a separate page* from the pages of text, indication of such address or endorsement will be separated from indication of the number of pages by "and". If such address and/or endorsement is written *on the same page* as a page of text, indication of such address or endorsement will be separated from indication of the number of pages by a comma. Thus, if a four-page letter, for example, has an endorsement on the top of the first page, above the text, it will be indicated "4p., end."; if the endorsement is on a separate, fifth page, it will be indicated "4p. and end." In other words, readers should always assume that the use of a comma in this instance means that the additional portion of the item is substantively, but not physically, distinct from what procedes the comma; use of the word "and" instead of a comma means that the additional portion is both substantively *and* physically distinct.

When there is either an address for Darwin's correspondent or an endorsement, it is reproduced fully in brackets following the indication of its existance. As there is such a large number of misspellings in these addresses and endorsements, "*sic*" was not employed to indicate them. Readers may assume a misspelling to have been made by Darwin or his correspondent. No attempt was made to identify with certainty the handwriting of endorsements, although those endorsements provided in the calendar are usually in the hands of Darwin's correspondents. As in the case of the date for each item, editorial conventions and textual devices used in the descriptive part of each entry are discussed below in the special sections on such matters.

The third part of each entry contains the abstract—the real meat of the item. Radoff claims that "the question of how full the abstract should be has long been the *bête noire* of calendarers,"[37] and no statement about calendaring could be truer. After much consideration,

---

36. Although it is somewhat out of fashion to give the dimensions—the argument against inclusion being that, given the disparities of different types of handwriting, size is a poor indicator of length of text—measurements are given in this calendar because of two special factors: 1) most of the letters in the calendar are in Darwin's hand, so there is an uncommonly good relationship between dimensions and length of text; and 2) the dimensions of Darwin's stationery in some cases can be used to corroborate a determination of a date for an undated letter. For example, a claim that an undated letter on stationery measuring eight by five inches was written in, say, 1860, will be corroborated if other Darwin letters known to have been written around this same date also are written on stationery of the same size. This is by no means a hard and fast rule, however; Darwin apparently reverted to leftover scraps and remainders of old stationery on occasion.

37. Radoff, "A Guide to Practical Calendaring," *op. cit.*, note 30, 134.

reasonably full abstracts were deemed desirable, but a price—namely the deletion of all textual matter which is published faithfully elsewhere and the rendering of the remaining textual matter in a complex way which is not exceptionally easy to read—had to be paid for this comprehensiveness, so that the length of the calendar would not be prohibitive. Accordingly, if it is known that the item or any part of it has been published, a simple reference to the location of the printed text will be made in lieu of an abstract of this portion of text. As the completeness and the accuracy of such printed texts varies considerably, however, a series of standardized phrases are employed in the references to the locations of the printed texts, and substantive corrections are supplied. These standardized phrases—arbitrarily designated as "printing conventions"—are designed to convey a rough idea of the accuracy and the completeness of the printed texts; they are discussed below in a special section. The substantive corrections—including abstracts of substantive deleted portions—follow reference to the location of the printed text. Corrections are keyed into the printed text by reference to the page and line of printed text being corrected. If necessary to avoid ambiguity, the exact location within the given line also is provided. When there is other material on the same printed page as a letter which is being corrected, and a correction is indicated for, say, line 15, the reader should begin counting lines only where the text of the incorrect letter begins: in other words, in this case "line 15" means "line 15 of the letter being corrected," not "line 15 from the top of the page."

If no published version of any part of the item is known, an abstract of the entire text is provided. The abstracts treat every topic in the original texts; topics are usually (but not always) presented in the order in which Darwin presented them. Each abstract consists of a series of statements (not necessarily whole sentences) connected by semicolons. When a clause begins with a singular indicative verb (e.g. "asks") and no subject of the verb is given, Darwin is implied as the subject. When the verb is not singular and indicative (e.g. "ask"), it should be read as a command to the recipient. Substantial quotation is employed. Superscript arabic numerals in this and the preceding parts of each entry indicate substantial editorial additions and refer to the correspondingly numbered notes in the fourth part of the entry. Other editorial conventions and textual devices employed in this part of each entry are discussed below in special sections.

The fourth and final part of each entry consists of annotations to the preceding three parts. An effort was made to identify every reference to persons, places, things, publications, and events; if any of these is not identified clearly in an entry, the reader can assume that research upon it proved unsuccessful. Annotations contain the minimum of explanation or interpretation, for this is not the responsibility of an editor; primary objectives of the calendar annotations were identification and description.

Biographical information is not provided in the annotations to each entry, as this would have necessitated a great many biographical notes for each person, one for each entry in which he or she is mentioned. Instead, biographical information is provided in the biographical notes following the calendar.

Numbering of the annotations begins afresh for each entry. Sometimes one note (usually the first one) relates to more than one location in the entry, usually because a single annotation can explain how both the date and the recipient of the item were determined in the absence of explicit information in the original. To save space, and to avoid an excessive number of annotations, certain information such as Christian names and titles of books are provided in the abstract rather than in the notes wherever this information can be worked into the abstract reasonably smoothly. Use of the word "perhaps" in a note indicates that the material in the note is conjectural.

To some degree, the letters annotate each other, since there is often a run of letters on the

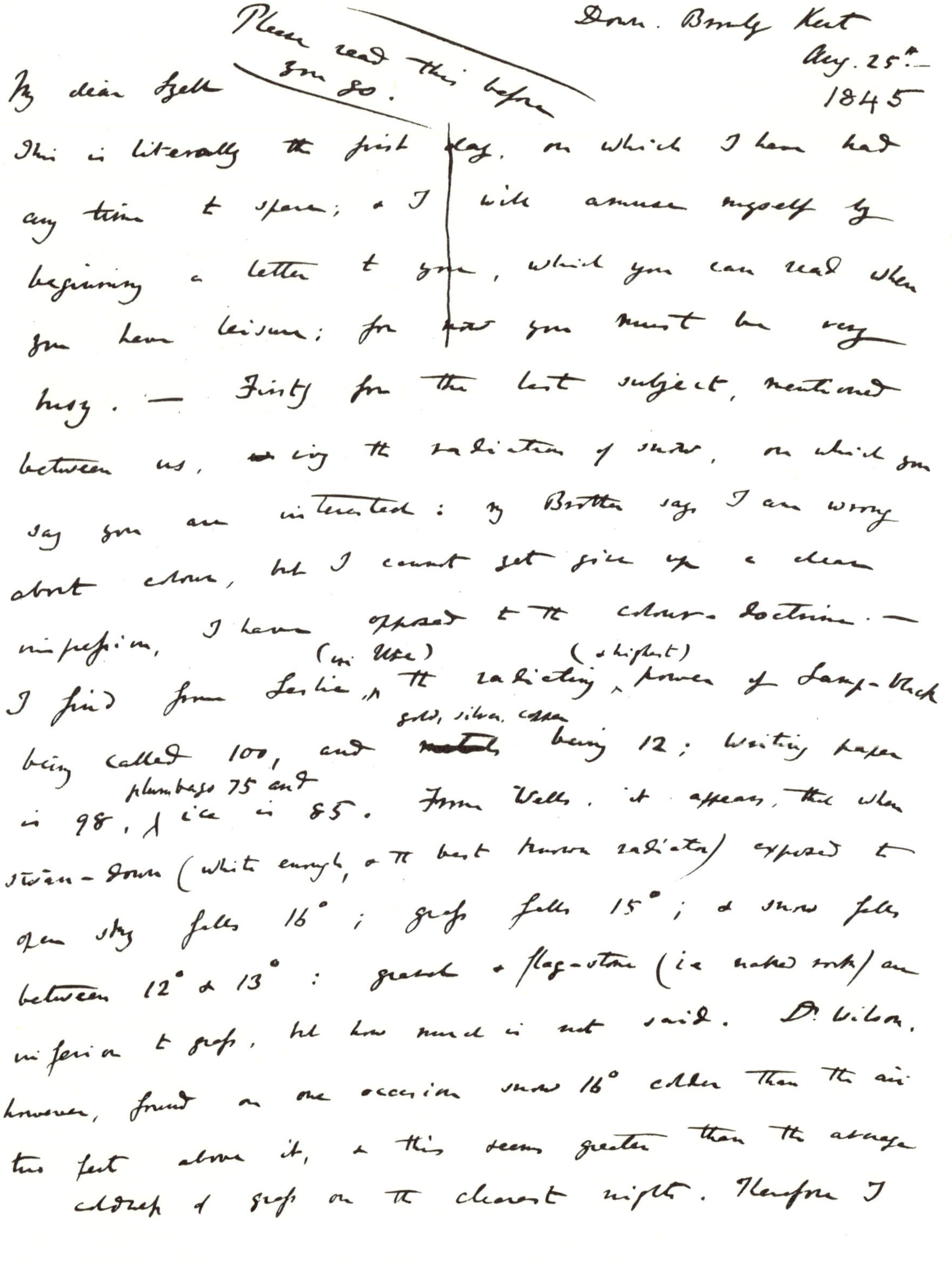

Please read this before you go.

Down. Bromley Kent
Aug. 25th—
1845

My dear Lyell

This is literally the first day on which I have had any time to spare; & I will amuse myself by beginning a letter to you, which you can read when you have leisure; for now you must be very busy.— Firstly for the last subject, mentioned between us, namely the radiation of snow, on which you say you are interested: my Brother says I am wrong about colour, but I cannot yet give up a clear impression, I have opposed to the colour-doctrine.— I find from Leslie, (in Ure) that the (slightest) radiating power of Lamp-black being called 100, and gold, silver, copper being 12; Writing paper is 98, plumbago 75 and ice is 85. From Wells it appears, that when swan-down (white enough, & the best known radiator) exposed to open sky falls 16°; grass falls 15°; & snow falls between 12° & 13°: gravel & flag-stone (i.e. naked rocks) are inferior to grass, but how much is not said. Dr. Wilson, however, found on one occasion snow 16° colder than the air two feet above it, & this seems greater than the average coldness of grass on the clearest night. Therefore I

Figure 3. First page of letter to Lyell, 25 August 1845. Note the reference to William Charles Wells's *Essay on Dew*, beginning eight lines from the bottom of the page. (American Philosophical Society)

same subject. Cross-references between two or more letters in which the same subject is discussed are kept to a minimum. Generally, those letters which are obviously part of a series of exchanges between Darwin and a given correspondent on some *particular* and *very definite* topic are referred only to the *next earlier letter* on the same topic. This allows readers to trace an exchange on a given topic backwards through time, once the last letter of the series is found.

Should readers wish at this time to see an example of the layout of the four parts of each entry, a good letter for study is that to Lyell dated August 25, [1845]; its entry is on page 18 below, and the first page of the manuscript letter is reproduced as Figure 3. This letter not only provides an example of the four parts of each entry but also allows readers to compare a typical calendar abstract against the original manuscript, thereby getting some notion of the thoroughness and reliability of the abstracts.

Additional discussions of minutiae of editorial policy appear at the head of the list of acronyms, abbreviations, and short titles given below, as well as at the head of the biographical notes following the calendar. Finally, in those few locations where editorial additions might easily be interpreted as part of the original item, the initials of the editor ("PTC") have been added to identify the material as editorially added.

*Editorial Conventions*

A calendar, unlike a letterpress edition, provides comprehensive faithfulness to the original manuscript only in certain portions of each entry. This imposes upon the calendarer the additional preliminary editorial chore of indicating precisely which parts of each entry are direct quotations from the original and which are not. In this calendar, an attempt has been made to preserve in each entry as much as possible of the original manuscript.

Subject to the editorial conventions to be outlined below, the following portions of each entry are to be considered as exact transcriptions from the original Darwin letter: the date (i.e. the first part of the entry), including watermarks, postmarks, and endorsements; the name of Darwin's correspondent (i.e. the first element of the first section of the description); Darwin's address (i.e. the second element of the first section of the description), except where the address was a variant Down address, in which case the shorthand form described earlier is used; the endorsement and the address of Darwin's correspondent, if these are provided in the original; those portions of the abstract which are enclosed in inverted commas (hereafter referred to as "quoted portions"); and the factual data (dates, names, titles of books, etc.) in those portions of the abstract *not* enclosed in inverted commas (i.e. "unquoted portions"). Only the earliest decipherable postmark is provided, and only that portion of a watermark indicating the date is given.

In each of these exactly-transcribed locations—with the exception of addresses, endorsements, postmarks, and watermarks, which are discussed below—the following rules apply: if the word, phrase, or clause is doubtful, the passage in doubt is enclosed in brackets and *followed* by a question mark ("?"). When more than one reading is plausible, the several readings are all enclosed in a single set of brackets, and each reading is followed by its own question mark (e.g. "[these? those?]"). Editorial additions are enclosed in brackets *without* a question mark. If the editorial addition is conjectural, it will be enclosed in brackets and preceded by a question mark. Deletions from the text are denoted by ellipsis points ("..."). Passages in parentheses are parenthetical matter from the original. See the section of textual devices below for a graphic representation of these rules.

Exception to these rules must be made in the case of addresses, endorsements, postmarks, and watermarks because, although they are faithful transcriptions of the original manuscript, these elements are always enclosed within brackets, as a matter of course, in order to set them off from contiguous matter of a different nature. To be specific, watermarks and postmarks are given with the date provided by Darwin; they are enclosed in brackets to prevent their being confused with the date which Darwin himself wrote. Similarly, addresses and endorsements are inserted as part of the description of the item; they are bracketed to set them off from this descriptive matter, the latter having been added editorially. In these special cases, matter in *parentheses* consists of either portions which are added by the calendarer, are dubious readings, or are found in parentheses in the original; which of these alternatives applies in each case is either clear from the context or is explained by a note.[38] Also, in these places where the bracket rules are exceptional, deletions are still denoted by ellipsis points and question marks preceding and following passages have the same meaning as in other locations. In the address for and the endorsement by Darwin's correspondent, as well as in the postmark, transition from one line of manuscript to the next is indicated by the slash ("/"); these transitions are not noted in any other portion of each entry.

In *all* locations, titles of office, such as "Mr.", "Mrs.", "Capt.", or Sir", are omitted except in cases in which Darwin gave such titles in the original and the title is or may be important in identifying the person to whom it refers. Also, occasionally a title of office (esp. titles of rank or royalty such as "Captain", "Colonel", or "Lady") will be retained to set endpoints for an otherwise undated letter by reference to the years during which the person had the title. (This only works, of course, if the person in question was elevated to a higher rank at a later time; otherwise, one might keep the same title for life.)

Names of persons are given fully enough in each location in which they appear so that a positive identification can be made. Those parts of the full name not provided by Darwin are, of course, enclosed in brackets. Sometimes, when necessary to distinguish between two persons with the same name (e.g. father and son), a title of honor or other distinguishing appellation will be annexed in brackets to the name, even if the title had not been conferred at the time the letter was written. An example of such an anachronistic usage is the addition of "Baron Avebury" to the name of John William Lubbock the younger (1834-1913) whenever it appears in letters written before he became a baron. Another anachronistic practice is the inclusion of a woman's married name before the date of her marriage (e.g. "Henrietta Emma Darwin Litchfield" before her marriage to Litchfield). In the case of multiple appearances of the same person's name in a single letter, the full name will be given only at the first encounter; succeeding references will be to surname only.

In the quoted portions of the abstract, Darwin's marginal notes, postscripts, stray marks, and marginal additions are inserted into the text of the letter without comment if the proper and precise location for them in the text has been indicated by Darwin and is readily identifiable. Otherwise, editorial comments describe roughly where such passages appear in the manuscript. In contrast, extraneous marks made by unknown third parties or by autograph dealers, or numbers and names written by Francis Darwin as part of his compilation scheme for *Life and Letters* (e.g. the "Denny 5" at the top of Fig. 2a) are not recorded.[39] In some cases, deciding whether a stray mark was an endorsement or an extraneous mark of some later handler was very difficult, and a subjective decision had to be made; at all times, an effort was made to be as systematic as possible.

---

38. An example of an *editorial addition* is the "(forwarded to)" in the address for Herbert in Darwin to Herbert, June, 1832, pp. 1-2 below. An example of a *dubious reading* is the "(S?)" in the address for Herbert in Darwin to Herbert, [September 13, 1828], p. 1 below. An example of *material appearing in parentheses in the original* is Lyell's "(105)" in his endorsement of Darwin to Lyell, [December (?19), 1837], p. 4 below.

39. Such marks by Francis Darwin are usually readily distinguishable because they often are written in a distinctive purple ink. Francis Darwin refers to these marks himself in a letter to Léo Abram Errera when he says: "Please excuse the numbers of reference with which I have marked the letters [from Charles Darwin to Errera, which Errera lent to Francis for use in the *Life and Letters*]." (F. Darwin to Errera, [October 25, 1882], APS; see appendix below.)

Darwin's cancellations (i.e. strike-outs) generally are omitted silently from the quoted portions, since they are usually trivial. In those few cases where a cancellation is retained, it is enclosed in brackets with an editorial comment indicating that it is a cancellation. Passages inserted between lines by Darwin are inserted silently into the text whenever Darwin indicated that they should be inserted.

The original punctuation, grammar, capitalization, and orthography are retained in all passages which are transcribed exactly from the original, with the following exceptions. Superfluous dashes at the close of sentences are usually deleted silently, although a few are kept to retain the Victorian flavor. Colons at the ends of abbreviations are changed to periods. Raised or superscript letters, such as were often found at the ends of abbreviations in nineteenth century correspondence, are silently lowered. Ligatures are silently expanded to two letters, and the elongated "s" is modernized silently. The ampersand ("&") and forms based upon it, such as "&c." for "et cetera", are retained. Other abbreviations in the original are usually expanded, with the added portions enclosed in brackets; in such cases of expansion, the period at the end of the abbreviation is silently deleted.

These few rules help keep the quoted portions as close to the original as is reasonably possible. The reason for such excessive faithfulness to the manuscript is that, since Darwin's penmanship is so poor, a great amount of deviation from the original might make it difficult for scholars unfamiliar with Darwin's hand to trace back from the choppy text of the abstract to either the original manuscript or a facsimile of it.[40] Such difficulty would be compounded, of course, if the scholar could see only a photocopy or microfilm, since—unavoidably—such facsimiles are often somewhat less readable than the already difficult manuscript. In some cases, however, such meticulous faithfulness to the original forces deviations from standard practices (e.g. titles of books not underlined). These are allowed to stand without comment where their meaning is not ambiguous; if there is any question as to Darwin's meaning (e.g. if the word "Origin" may or may not refer to the *Origin of Species*), any information which could be found to resolve the ambiguity is added in brackets or in a note.

Regarding unquoted portions of the abstract, the original phraseology of the item is often employed without comment, in an effort to minimize the distortion that inevitably results from summarization. In such unquoted portions, spelling, punctuation, and capitalization are modernized and converted to American forms.

Examples of the application of many of these editorial conventions can be observed by comparison of the following three versions of the letter from Darwin to Lyell, Friday, [June 25, 1858]: 1) the calendar entry, page 56; 2) the first five pages of manuscript for this letter, reproduced as Figure 4; and 3) the printed text on pages 117-18 of volume 2 of *Life and Letters.* This will show the relative thoroughness with which a printed text has been corrected in the abstract, the relative faithfulness with which the quoted portions of the abstract follow the original text, and the relative amount of condensation which occurs when passages of the original are reduced to unquoted portions of the abstract.

*Textual Devices*

| | |
|---|---|
| . . . | Portion of original text is deleted. |
| / | End of a manuscript line; used only for postmarks, and for addresses of and endorsements by Darwin's correspondents. |

---

40. In addition, despite the many errors and other indications of hurriedness in Darwin's letters, the many corrections in the letters indicate that the meticulous Darwin paid reasonably close attention to details of spelling, grammar, punctuation, and the like. A good illustration of this trait appears in a letter to Lyell in which Darwin tells the geologist that he had misspelled "Van Diemen's Land [i.e. Tasmania]" in a manuscript (Darwin to Lyell, [August 2, 1845], APS, calendar listing below, pp. 17-18; this part of the letter is not brought out in the calendar entry.)

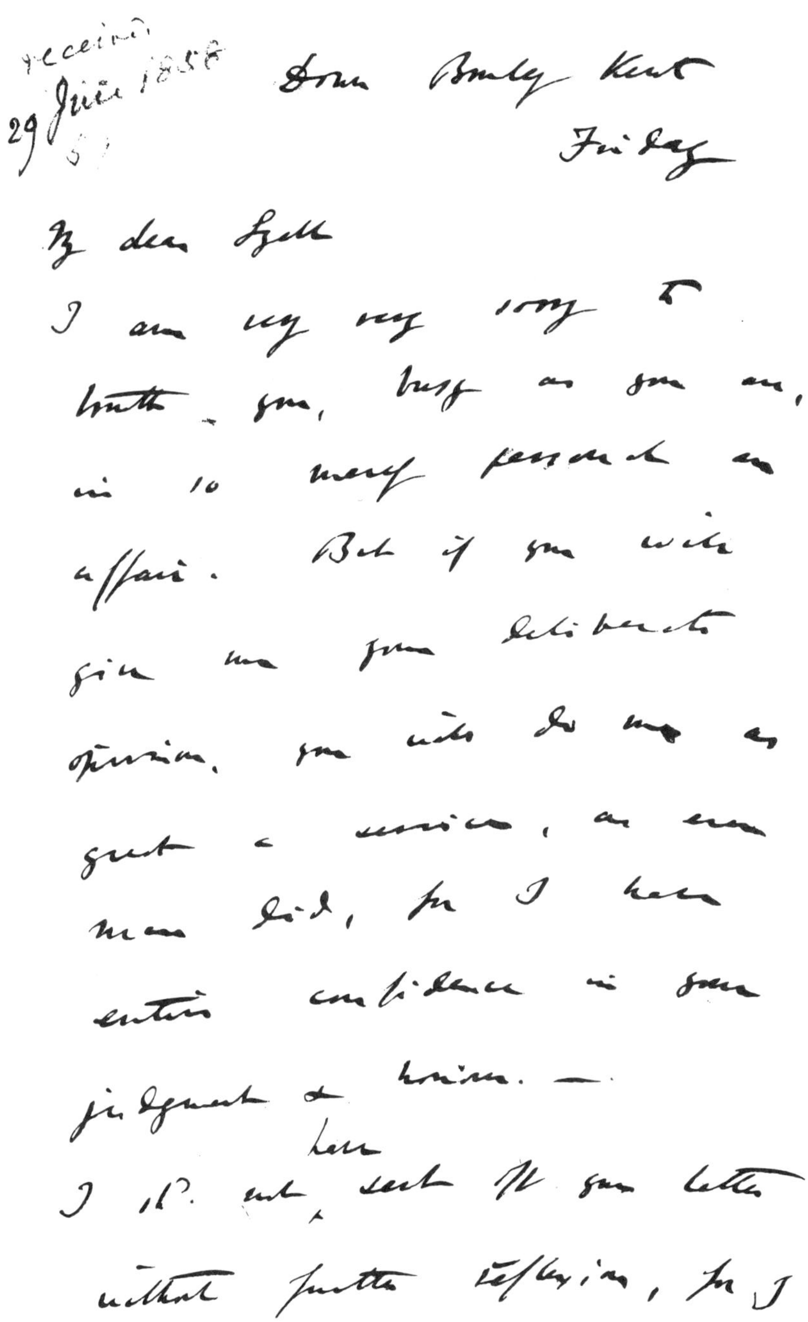

received
29 June 1858

Down Bromley Kent
Friday

My dear Lyell
I am very very sorry to trouble you, busy as you are, in so merely personal an affair. But if you will give me your deliberate opinion, you will do me as great a service, as ever man did, for I have entire confidence in your judgment & honour. —
I sh<sup></sup>. not have sent off your letter without further reflexion, for I

Figure 4. First five pages of letter to Lyell, Friday, [June 25, 1858]. In this famous letter Darwin argues the propriety of publishing his theory of natural selection after having received an essay by Alfred Russel Wallace containing a summary of the same views. The printed version of this letter (*Life and Letters,* II, 117-18) does not contain the sentence beginning three lines from the bottom of the fourth page, which reads "I do not in least believe that . . . [Wallace] originated his views from anything which I wrote to him"; this is a representative example of the occasional substantive deletions in printed versions of Darwin's letters. (American Philosophical Society)

am at present quite upset,
but write now to get
subject for time out of
mind. But I confess it
never did occur to me,
as it ought, that Wallace
could have made any
use of your letter.
There is nothing in Wallace's sketch
which is not written out much
fuller in my sketch copied
in 1844, & read by Hooker
some dozen years ago. About
a year ago I sent a short

sketch, of which I have copy, of my views (owing to correspondence
on several points) to Asa Gray,
so that I could most truly
say & prove that I take
nothing from Wallace. I
should be extremely glad now
to publish a sketch of
my general views in about
a dozen pages or so. But
I cannot persuade myself
that I can do so honourably.
Wallace says nothing about
publication, & I enclose
his letter.— But as I had

not intended to publish any sketch, can I do so honourably because Wallace has sent me an outline of his doctrine? — I would far rather burn my whole book than that he or any man should think that I had behaved in a paltry spirit. Do you not think his having sent me this sketch ties my hands? I do not in least believe that he originated his views from anything

59

which I wrote to him. If I could honourably publish I would state that I was induced now to publish a sketch (& I should be very glad to be permitted to say to follow your advice long ago given) from Wallace having sent me an outline of my general conclusions. — We differ only, that I was led to my views from what artificial selection has done for domestic animals.

| | |
|---|---|
| [ ] | Editorial addition. |
| [ ?] | Conjectural reading. |
| [? ] | Conjectural editorial addition |
| ( ) | Parenthetical marks found in original text, *except* when used in conjunction with addresses, endorsements, postmarks, and watermarks. (See text for explanation of these exceptions.) |
| [CD brackets] | Bracketed entry which immediately precedes this textual device is a bracketed parenthetical remark in the original text. |

*Descriptive Abbreviations*

Descriptive abbreviations always indicate three things: the type of document being calendared; whether or not the document is in the hand of the author of the document; and the presence or absence of the author's signature. Sometimes they also give additional information, such as indication that the signature is in the form of initials only. These abbreviations are more-or-less standardized among archivists, and they are discussed elsewhere in considerable detail,[41] so only a brief explanation of them is provided below. Those unfamiliar with such abbreviations may consult the more extended discussion published elsewhere for details.

As used in this calendar, descriptive abbreviations consist of a combination of one or more *basic abbreviations* (usually each basic abbreviation is a single letter) into a *composite abbreviation* which fully describes the document being calendared. The basic abbreviations used are the following:

| | |
|---|---|
| A | Autograph; the item is written in the hand of the author. |
| add. | Address; the address of the recipient has been provided. |
| by init. | By initials; the signature consists only of the author's initials. |
| D | Document; the item is a writing of some sort, but is not a letter or a note, and it is in reasonably final form. |
| end. | Endorsement; a brief note written on the item by the recipient. |
| L | Letter; the item is a written communcation of some length. |
| N | Note; the item is a very short written communcation. |
| p.c. | Postal card; the item has been written upon a standard postal card. |
| S | Signed; the item has been signed by its author. |
| sketch | Sketch; part of the item is a drawing in the hand of the author. |
| T | Typed; the item has been typed rather than written. |

These basic forms can be combined to form composite abbreviations which describe the item being calendared. It would be foolish to list all the possible permutations of the basic forms here, especially since the many possible composite abbreviations can be suggested well by a few examples, such as the following: "ALS" indicates a holograph letter signed by the author; "LS" indicates a letter signed by the author but written in the hand of an amanuensis; "AN on p.c., S by init." indicates a holograph note written on a postal card and initialed, but not signed in full, by the author. The other possible combinations should follow logically from these examples, especially if readers keep in mind the following rule of thumb: assume that the author or composer of the item had nothing to do with the actual writing of the item unless indicated otherwise.

41. Radoff, "A Guide to Practical Calendaring," *op. cit.*, note 30, 136-40

*Printing Conventions*

Since many of the letters in the calendar have been published in whole or in part, and since both the completeness and the accuracy of these printed versions vary considerably, the following conventional phrases have been employed to indicate roughly how well the printed versions resemble the original letters. The conventions were devised so that—in the absence of contrary indications—the reader may assume that only a part of the item being calendared has been printed. The reader may further assume that no alterations of the original text have been made in the printed version; if, on the other hand, alterations *are* indicated, the reader may assume that these changes are of a serious nature, unless a contrary indication is provided. The conventions are listed in descending order of resemblance to the original (assuming changes to be more odious than omissions):

| | |
|---|---|
| Printed in facsimile | The manuscript letter is photographically reproduced. |
| Printed in full | The item is fully and faithfully transcribed in the printed version. |
| Printed in full, with minor changes | The item is fully transcribed in the printed version, but changes in such details as punctuation, spelling, capitalization, inclusion or deletion of articles of speech, or ungrammatical usages have been made without comment; in some cases, minor words may be incorrectly transcribed, but no alteration of Darwin's intended meaning has resulted. |
| Printed, with minor changes and minor omissions | The item is almost fully transcribed in the printed version; these portions which are missing are not significant and do not concern a subject not mentioned in the printed version; those portions which are printed contain changes in the text as described in "Printed in full, with minor changes" above. |
| Printed | The item is not fully transcribed in the printed version; those portions which are missing are significant and/or concern a subject not mentioned in the printed version; those portions which are printed are fully and faithfully transcribed. |
| Printed, with minor changes | The item is not fully transcribed: missing portions are significant and/or concern a subject not mentioned in the printed portion; those portions which are printed contain changes in the transcript as described in "Printed in full, with minor changes" above. |
| Printed in full, with changes | The item is fully transcribed, but serious changes have been made which may affect the meaning of all or part of the text. |
| Printed, with changes and minor omissions | The item is almost fully transcribed; missing portions are not significant and do not concern a subject not mentioned in the printed portion; those portions which are printed contain changes in the transcript as described in "Printed in full, with changes" above. |
| Printed, with changes | The item is not fully transcribed; missing portions are significant and/or concern a subject not mentioned in the printed portion; those portions which are printed contain changes in the transcript as described in "Printed in full, with changes" above. |

Corrections provided in the abstracts are designed to rectify the changes and major omissions. When a major omission has been rectified in the abstract, the printed version is described as having only minor omissions.

# Acronyms, Abbreviations, and Short Titles

Abbreviations of periodical titles not listed below may be found in the *World List of Scientific Periodicals,* 1900-1960, fourth edition. Abbreviations of titles not found in the *World List* were derived by the editor by using the system of abbreviation described in the *World List*; the resultant abbreviated titles are listed below. This rather unorthodox approach to abbreviating periodicals in an American publication (i.e. using an English source for abbreviations rather than, say, the *Union List of Serials*) was employed because the *World List,* unlike other sources, includes many of the more obscure titles cited by Darwin, and also because the *World List* abbreviations seem to the author to be the most systematic, consistent, and understandable of any yet devised.

With exceptions noted below, Darwin titles are shortened to the forms used as headings in Part 2 of R.B. Freeman, *The Works of Charles Darwin: An Annotated Bibliographical Handlist,* [first edition] (London: Dawsons of Pall Mall, 1965). For each Darwin work cited, the short title and the year of publication only are given, except in cases in which more than one issue of that title appear in the given year; in such cases, the number of the issue used, as found in the Freeman *Handlist,* is also provided. In places where only the short title and the year appear, and Freeman indicates that more than one issue of that title appeared in that year, the reader may assume that what is said applied to all of these issues or variants.

Common abbreviations, such as "univ." for "university", are omitted, of course.

AAAS–American Association for the Advancement of Science.

*Abstr. Pap. Communicated R. Soc. Lond.–Abstracts of Papers Communicated to the Royal Society of London* (usually considered to be volumes five and six of the *Proceedings,* 1843 to 1854).

APS–American Philosophical Society, Held at Philadelphia, for Promoting Useful Knowledge.

B.A.A.S.–British Association for the Advancement of Science.

Barlow, ed., *Autobiography*–Charles Darwin, *The Autobiography of Charles Darwin, 1809-1882, with Original Omissions Restored,* edited with appendix and notes by . . . Nora Barlow [*Freeman* 371] (London: Collins, 1958).

*Brit. for. med.-chir. Rev.–British and Foreign Medico-Chirurgical Review.*

CD–Charles Robert Darwin, 1809-1882.

*Can. Naturalist–Canadian Naturalist and Quarterly Journal of Science, with Proceedings of the Natural History Society of Montreal.*

corr.–correspondent.

*DAB–Dictionary of American Biography.*

Darwin, *Coral Reefs–The Structure and Distribution of Coral Reefs, Being the First Part of the Geology of the Voyage of the Beagle* . . . (London: Smith, Elder and Co., 1842).

Darwin, *Fossil Balanidae*–Charles Darwin, *A Monograph on the Fossil Balanidae and Verrucidae of Great Britain* [*Freeman* 105, v. 2] (London: Palaeontographical Society, 1854).

Darwin, *Fossil Lepadidae*–Charles Darwin, *A Monograph on the Fossil Lepadidae, or, Pedunculated Cirripedes of Great Britain* [*Freeman* 105, v. 1] (London: Palaeontographical Society, 1851).

Darwin, *Origin* (year)–Charles Darwin, *On the Origin of Species by Means of Natural Selection, or the Preservation of Favoured Races in the Struggle for Life,* edition and issue determined by year of publication and, if necessary, by *Freeman* number.

Darwin, *Recent Balanidae*—Charles Darwin, *A Monograph on the Sub-class Cirripedia, with Figures of All the Species: The Balanidae, (or Sessile Cirripedes); The Verrucidae, etc., etc., etc.* [*Freeman* 103, v. 2] (London: The Ray Society, 1854).

Darwin, *Recent Lepadidae*—Charles Darwin, *A Monograph on the Sub-class Cirripedia, with Figures of All the Species: The Lepadidae; or, Pedunculated Cirripedes* [*Freeman* 103, v. 1] (London: The Ray Society, 1851).

Darwin, *South America*—Charles Darwin, *Geological Observations on South America, Being the Third Part of the Geology of the Voyage of the Beagle...* (London: Smith, Elder and Co., 1846).

Darwin, *Volcanic Islands*—Charles Darwin, *Geological Observations on the Volcanic Islands, Visited during the Voyage of H.M.S. Beagle, Together with Some Brief Notices on the Geology of Australia and the Cape of Good Hope, Being the Second Part of the Geology of the Voyage of the Beagle...* (London: Smith, Elder and Co., 1844).

*Darwin and Henslow*—Nora Barlow, ed., *Darwin and Henslow: The Growth of an Idea. Letters, 1831-1860* (London: Bentham-Moxon Trust, John Murray, 1967).

*DNB*—*Dictionary of National Biography.*

*DSB*—*Dictionary of Scientific Biography.*

*Edinb. J. nat. geogrl Sci.*—*Edinburgh Journal of Natural and Geographical Science.*

*Edinb. new phil. J.*—*Edinburgh New Philosophical Journal.*

*Edinb. Rev.*—*Edinburgh Review.*

*Emma Darwin*—H. E. Litchfield, ed., *Emma Darwin, Wife of Charles Darwin: A Century of Family Letters,* 2v., privately printed [*Freeman* 359] (Cambridge: Cambridge University Press, 1904).

FGS—Fellow of the Geological Society of London.

FLS—Fellow of the Linnean Society, London.

*Fortn. Rev.*—*Fortnightly Review.*

FRCP—Fellow of the Royal College of Physicians, London.

*Freeman*—R. B. Freeman, *The Works of Charles Darwin: An Annotated Bibliographical Handlist* (London: Dawsons of Pall Mall, 1965).

FRS—Fellow of the Royal Society of London.

FRSE—Fellow of the Royal Society of Edinburgh.

FZS—Fellow of the Zoological Society of London.

*Handlist of Darwin Papers*—*Handlist of Darwin Papers at the University Library Cambridge* (Cambridge: Cambridge University Press, 1960).

iss.—issue.

*J. R. geogrl Soc. Lond.*—*Journal of the Royal Geographical Society, London.*

*Larousse*—Pierre Larousse, *Grand Dictionnaire Universel du XIXe Siècle....*

*Life and Letters*—Francis Darwin, ed., *The Life and Letters of Charles Darwin, Including an Autobiographical Chapter,* 3v. [Either *Freeman* 346 (first edition, 1887) or *Freeman* 349 (seventh thousand revised, 1888)] (London: John Murray, 1887 or 1888). A simple *Life and Letters* citation, therefore, indicates that the citation is correct for *either* of the two editions listed above; compare this with the next entry below.

*Life and Letters* (seventh thousand revised, 1888)—Francis Darwin, ed., *The Life and Letters of Charles Darwin, Including an Autobiographical Chapter,* seventh thousand revised, 3v. [*Freeman* 349] (London: John Murray, 1888). A short-title citation of this type indicates a discrepancy in the pagination for the cited material between the first and the revised editions of the *Life and Letters*; the pages cited apply to the revised edition, while the same material can usually be found in the first edition a few pages later.

*Life of Lyell*–[K. M. H.] Lyell, ed., *Life, Letters, and Journals of Sir Charles Lyell, Bart.*, 2v. (London: John Murray, 1881).

*Life of Romanes*–[Ethel Duncan] Romanes, ed., *The Life and Letters of George John Romanes* (London, New York, and Bombay: Longmans, Green, and Co., 1896).

*Lyell: The Years to 1841*–Leonard G. Wilson, *Charles Lyell, The Years to 1841: The Revolution in Geology* (New Haven and London: Yale University Press, 1972).

*Mag. nat. Hist.*–*Magazine of Natural History* (ultimately merged with *Annals of Natural History* to become *Annals and Magazine of Natural History*).

*MEB*–Frederic Boase, *Modern English Biography.*

*More Letters*–Francis Darwin, and A. C. Seward, eds., *More Letters of Charles Darwin: A Record of His Work in a Series of Hitherto Unpublished Letters,* 2v. [*Freeman* 359] (London: John Murray, 1903).

MRCS–Member of the Royal College of Surgeons of England.

*Nat. Hist. Rev.*–*Natural History Review: A Quarterly Journal of Biological Science.*

*N. Br. Rev.*–*North British Review.*

Peckham, *Variorum Origin*–Morse Peckham, ed., *The Origin of Species by Charles Darwin: A Variorum Text* (Philadelphia: University of Pennsylvania Press, 1959).

pmk.–postmark.

*Poggendorff*–J. C. Poggendorff, ed., *Biographisch–Literarisches Handwörterbuch zur Geschichte der Exacten Wissenschaften.*

*Proc. geol. Soc. Lond.*–*Proceedings of the Geological Society of London* (preceded the *Quarterly Journal*).

*Proc. nat. Hist. Soc. Dubl.*–*Proceedings of the Natural History Society of Dublin.*

PTC–Material added by the compiler, P. Thomas Carroll.

*Q. Rev.*–*Quarterly Review.*

*Sat. Rev.*–*Saturday Review of Politics, Literature, Science and Art.*

*Silliman's J.*–*American Journal of Science* (known popularly as *Silliman's Journal of Science*).

Stauffer, ed., *CD's Nat. Selection*–Robert C. Stauffer, ed., *Charles Darwin's Natural Selection, Being the Second Part of His Big Species Book Written from 1856 to 1858* (Cambridge: Cambridge Univ. Press, 1975).

*Trans. geol. Soc. Lond.*–*Transactions of the Geological Society of London.*

Vorzimmer, *Reprint Catalogue*–Peter J. Vorzimmer, comp., *A Catalogue of the Darwin Reprint Collection at the Botany School Library, Cambridge* (Cambridge: unpublished mimeograph, 1963).

wmk.–watermark.

1 [1828 Sept. 13] Saturday Evening [pmk. Se 14/ 1828][1]

To [John Maurice] HERBERT; [Osmaston, near Derby]
ALS; 9 X $7\frac{1}{2}$; 4p., add. [(S?) Herbert Esqr/ Post Office/ Barmouth/ N. Wales]

Printed, with minor changes: Life and Letters, I, 172-73. p. 173, line 23, change "Yates" to "Yate". At end of letter is: "How is Buz & Bossy [sic]. I am afraid you yourself must be grown terribly bumptious: Direct to Shrewsbury: if there is any thing you want I can send it for you to Barmouth. Such as gloves &c &c &c".

1. Printed letter has erroneous date of Sept. 14, which was a Sunday.

2 [1828 October 3] Friday [pmk. Oc 4/ 1828; wmk. 1824]

To G [i.e. John Maurice] HERBERT; [pmk. Shrewsbury]
ALS; 9 X $7\frac{1}{4}$; 4p., add. [G. Herbert Esqr./ Court Calmore/ Welch Pool]

Obliged for "labours in the science [of entomology]"; saw [?Thomas] Butler who told of Herbert in Wales; Herbert's courage cooled since CD left Barmouth; chiding letter from [Charles Thomas] Whitley, answered humbly by CD, received mere "acknowledgement of my extreme candour" and another charge of idleness; supposes Herbert enjoys Montgomeryshire and delights "all the little dear female hearts"; CD enjoys successful Music Meeting in Derbyshire, also good shooting, "Entomological pursuits", and the "Miss Foxes[1] are very pleasant girls"; Herbert can give CD beetles & butterflies, which Butler says Herbert has, when Herbert is in Shrewsbury; CD "shall go up [?to Cambridge] early," but not by 10th; Butler goes next Tues.; P.S.: Find more beetles; get lady with "strong imagination" to procure beetle with "face so very dreadful"; forgets Herbert's Christian name, christens him "G" [see above].

1. The Misses Fox were the sisters of William Darwin Fox, CD's second cousin; the Music Meeting was held at the Fox residence. See "Darwin's Journal," 6.

3 [1831 September 9] Friday Evening [pmk. 10 SE 1831; wmk. 1830]

To Cha[rle]s WHITLEY; 17 Spring Gardens, London
ALS; 9 X $7\frac{1}{2}$; 4p., add. [Chas. Whitley Esqr./ Post Office/ Barmouth/ N. Wales], end.

Printed in facsimile: A Letter of Charles Darwin about Preparations for the Voyage of the Beagle, 1831 (Philadelphia: Friends of the Library, American Philosophical Society, 1971).

4 1832 June [pmk. (SE?) 30/ 1832; wmk. 1830]

To J[ohn] M[aurice] HERBERT; Botofogo Bay, Rio de Janero [sic]

ALS; 11 X 8 3/4; 4p., add. [J. M. Herbert Esqr./ Fellow of St. Johns Coll:/ Cambridge; (forwarded to) (W?) Maddy Esqr/ Moreton/ Near Hereford]

Printed, with minor changes: Life and Letters, I, 238-40. p. 239, line 32, after "contained", insert: "Tell [Charles Thomas] Whitley that I find my life on blue water...very pleasant,...an excellent time for reading; so quiet & comfortable, that you are not tempted to be idle." p. 240, line 2, after "reason...", insert: A short or stupid letter would end correspondence between some, "but old gentleman, you might as well try to cut your tailor as me"; letter from Herbert brings to CD "a thousand pleasant thoughts"; CD can picture Herbert "in the two extreme cases, of the dead March to Dolgelley & the bogtrotting Match with [?William] Selwyn." At end of letter is: P.S. "I have directed to you in a curious manner for fear of mistakes."

5 1833 June 2d. [pmk. Oc 2/ 1833; wmk. 1828]

To J[ohn] M[aurice] HERBERT; Maldonado, Rio Plata
ALS; 9 3/4 X 7 3/4; 4p., add. [J. M. Herbert Esqr/ Fellow of St John's Coll:/ Cambridge; (forwarded to) Lower Garthmyl/ Welshpool]

Printed, with minor changes: Life and Letters, I, 246-48. At end of letter is: please write again and remember CD to friends, including [Charles Thomas] Whitley; "Read [Francis Bond] Head's gallop[1] if you want an accurate account of this country." On first page of letter, written sideways, is: Has Herbert heard from F[?rederick] Watkins, [Jonathan Henry Lovett] Cameron or Matthews[?]; CD wrote to former many months ago, but no answer; address in future to be Valparaiso. On second and third pages of letter, written sideways, and perhaps not in CD's hand, is: "I have just met with the following quotation in the 'Sacred History of the World' taken from the Hereford!! Journal, November 1824.[2] 'Carnations have been engrafted on Fennel & for the first two or three years the flowers will be green: Likewise Peaches on a Mulberry, in which case the fruit will have a purple dye to the stone.' Were you the original & ingenious experimentalist? I think I have heard you argue that White Lies do no harm.-- Here are green Carnations & purple Peaches brought foreward [sic] to show the beneficence of Providence.-- When such evidence is proved false who will not become a Sceptic.-- Reflect--, if the author, what awful consequences may have been produced.--"

1. Sir Francis Bond Head, Rough Notes Taken during Some Rapid Journeys across the Pampas and among the Andes (London: n.p., 1826).
2. See: Sharon Turner, The Sacred History of the World, as Displayed in the Creation and Subsequent Events to the Deluge..., 3v. (London: Longman, Rees, Orme, Brown, Green, and Longman, 1832-1837), I, 111 and 111n; letter from "Ethelbert" in Hereford Journal..., November 24, 1824; and Richard Bradley, A General Treatise of Husbandry and Gardening..., 2v. (London: T. Woodward and J. Peale, 1726), II, 301. F. A. Milligan, Sydney Smith, and David Wilson have assisted me with this information.

6 [1836 October-November][1] Monday [wmk. 1832]

To Lieut[enant Charles] WILKES; 43 Grt. Marlborough St
ALS; $7\frac{1}{2}$ X $4\frac{1}{2}$; 1p. and add. [Lieut: Wilkes/ Long's/ Bond Street.--]

Is going into country for few weeks on Thursday; wishes to converse concerning Wilkes's "long...voyage"; unless Wilkes writes to contrary, CD "will call at Long's on Wednesday" between noon and 1 p.m.

1. CD did not return from the voyage of the Beagle until October 2, 1836 ("Darwin's Journal," 7). Wilkes was only in England from August to November, 1836 (Doris Esch Borthwick, "Outfitting the United States Exploring Expedition: Lieutenant Charles Wilkes' European Assignment, August-November, 1836," Proc. Am. phil. Soc., 109 [1965]: 159-72). The Great Marlborough Street address given by CD is that of his brother's flat, at which CD visited and roomed occasionally until leaving Cambridge and taking his own rooms at number 36 down the street on March 13, 1837 (Life and Letters, I, 277; Darwin and Henslow, 118 and 118n).

7 [ca. 1836-1837][1] Monday Evening

To the Master & Fellows [of] Caius College; no location
AL in third person; $4\frac{1}{2}$ X 3 3/4; 1p. and add. [The Master & Fellows/ Caius College]

"Mr Darwin presents his compliments to the Master & Fellows of Caius Coll. and is extremely sorry he is prevented by a previous engagement the honor [sic] of dining with them on Thursday.--"

1. Date determined by Sydney Smith, St. Catharine's College, Cambridge. The handwriting of this note appears to be in the hand of the young Darwin, and since this note shows no sign of being mailed--indeed seems to have been hand-delivered--it seems reasonable to conclude that CD was in Cambridge at the time it was written. This occurred in late 1836 and early 1837, right after CD returned to England from the Beagle voyage.

8 [1837 late September]

To Mr. [?Frederick or William] SHOBERL; [36 Great Marlborough St.]
ALS; 7 X $4\frac{1}{2}$; 3p., add. [Mr Shoberl/ Marlborough Sqr] (partially mutilated,

Obliged for document which CD "had full right to demand", even if unnecessary; sorry for inconvenience; will send completed MS. with woodcuts before night; will write to printer about "where to send the slips"; [Henry] Colburn will see revise corrected; shall go to Shrewsbury on Monday; gives printing details; thanks Shoberl and Colburn for aid on "this my first publication."[1]

1. CD's "first publication" was his Journal of Researches [Freeman 4] (1839), the only Darwin work published by Colburn. Date for this letter

derives from date CD finished the MS. of this work and when he left for Shrewsbury, shortly thereafter; see "Darwin's Journal," 7-8.

9 [1837 December (?19); end. Decr. 1837; pmk. DE 20/ 1837]

To Cha[rle]s LYELL; no location
ALS; 9 X 7½; 4p., add. [Chas. Lyell Esqr/ 16 Hart St.], end. [(105) Mr. Darwin on Coral islands thinly scattered over what area--Decr. 1837]

Does not know latitude limits of true coral islands in Pacific, but Bermuda an exception to general rule; discusses and describes coral islands and archipelagoes; "People's ideas of the Pacific are most false."; describes and sketches the "Corallian Sea" proposed by [Matthew] Flinders; thanks for books, but cannot read them for nearly a week, since still reading [Jean Baptiste Armand Louis Leonce] E[lie] de B[eaumont]; P.S. Tuesday night, longitudinal boundaries of Pacific coral cannot be given satisfactorily, since Dangerous or Low Archipelago and Corallian Sea are separated by "great volcanic band".

10 [1838] Aug. 9th. [end. 9 August 1838]

To [Charles] LYELL; 36 Grt. Marlbro' St.
ALS; 9 X 7½; 11p. and end. [(1) Mr Darwin on Elements & Glen Roy/ 9 August 1838]

Printed, with minor changes and omissions: Life and Letters, I, 291-95. p. 292, line 14, change "...." to "[John] Phillips will not surely go on saying that the metamorphic schists are disintegrated granite re-deposited."; p. 292, line 16, change "-----" to "Phillips"; p. 292, line 28, insert: CD visited Wednesday night, two days after Lyell left, thinking Lyell would come to London after Crag expedition; almost wrote from Shrewsbury; p. 294, line 19, change "-----'s" to "[Rev. Frederick William] Hope's"; p. 295, line 5, change "-----" to "Jones"; p. 295, line 13, insert: CD wants daytime barometer readings made at Leith on July 5, published by Brewster; also wants altitude of Lochs Tay, Dochart, Tyndrum, and Tulla;[1] p. 295, insert: "P.S. I have seen [Robert] Fitzroy, who has bought your book.[2] He looked rather black at the preface...but then came smooth again. I never cease wondering at his character,...full of good...traits but spoiled by such an unlucky temper.-- Some part of... his brain wants mending...."

1. Lyell apparently complied; see Darwin, "Observations on the Parallel Roads of Glen Roy,..." Phil. Trans. R. Soc., 129 (1839), 54.
2. Lyell, Elements of Geology (London: John Murray, 1838). The preface declares that the publication of Darwin's Journal of Researches [Freeman 4] (1839) has been delayed, "to the great regret of the scientific world," by the failure of Fitz Roy to complete the companion volumes to it.

11 [1838] September 13th Friday Night [end. 1838; pmk. SP15/ 1833]

To Charles LYELL; no location
ALS; 12 3/4 X 8; 4p., add. [Charles Lyell Esqr Junr./ Kinnordy/ Kerrimuir/ North Britain], end. [Darwin 1838 on tortuosity of parallel bands of Elevation & subsidence--/ (2)]

Printed, with minor changes: Life and Letters, I, 295-98.
p. 296, line 2, insert: discussion concerning "unfortunate letter" of Governor [Henry] Prescott, a lost letter of CD, an "official document", an apparently abused frank from Lord [Spencer Joshua Alwyne Compton] Northampton, and a proposal to the Council of the Royal Society of London; p. 296, line 4, after "begin about.", insert: thanks for holiday invitation, but CD pledged to visit uncle and home for fortnight at end of October,[1] and "till then I must not give myself even a days holidays."; p. 297, line 22, change "block" to "Polar"; p. 297, line 24, insert: "Sir D[avid?] B[rewster?] ...communicated some information so useful, that I have written to him again."; gives details on "Winters Bark", parrots, geographical distribution in South America and in Falkland Islands, Port Famine; sent Lyell's letter to "Dr. Richardson at Portsmouth"; unhappy that Lyell says he will be away until end of November, hopes "something will bring you back before then."; p. 298, line 4, insert derogatory remarks about Babbage and his calculator; p. 298, line 6, insert: regarding marsupials, CD saw only abstract of [Henry Marie Ducrotay de] Blainville's paper;[2] [Richard] Owen a better authority than Blainville, who is superficial;

Owen says internal process in Stonesfield Jaws is confined to marsupial mammals[3] and "talks of [Ornithorhynchus, the duckbill] leading off into the reptiles...[therefore] some reptiles formerly might have approached nearer to the Mammalian type, than...existing ones now do."; Elements [of Geology, Lyell's new book] must be selling well, requires "hard reading" and thus does not shirk its subject, as do two of [John Frederick William] Herschel's treatises; [Edward] Charlesworth is annoyed because Lyell did not quote him more; Charlesworth is to be pitied for many reasons; Zoological Society is giving up Associate Secretary's place; [John] Gould's case of Water-wagtails does not hold.

1. CD became engaged during this "pledged" holiday; see "Darwin's Journal," 8.
2. Probably either "Doutes sur le Prétendu Didelphe Fossile de Stonesfield," C. r. hebd. Séanc. Acad. Sci., Paris, 7 (1838), 402-18, or "Nouveaux Doutes sur le Prétendu Didelphis de Stonesfield," ibid., 727-36 and 749-51.
3. See Richard Owen, "Observations on the Fossils Representing the Thylacotherium Prevostii, Valenciennes,..." Trans. geol. Soc. Lond., 6 (1842), 47-65; for fuller account, see idem, A History of British Fossil Mammals and Birds (London: John Van Voorst, 1846), 29-57.

12 [1838 November] 13th [i.e. 12th] Monday [end. Novr. 1838; pmk. NO 13/ 1838; wmk. 1834]

To Cha[rle]s LYELL; Shrewsbury
ALS; 9 X 7¼; 3p. and add. [Chas. Lyell Esqr/ 16 Hart St./ Bloomsbury], end. [Mr Darwin/ Novr. 1838]

Printed in full, with minor changes: Emma Darwin, I, 413-14.

13 1838 Decr. 20th [end. March 1, 1839, wmk. 1836]

To The Secretary of the American Philosophical Society, [Franklin BACHE]; Geological Society of London/ Somerset House
Printed L, filled in in ms. (not CD's hand), S by CD; 15 X 9; 1p. and add. [The Secretary of the American Philosophical Society], end. [A.P.S.] Stated Meeting/ March 1, 1839, Read]

Routine thanks for Proceedings of the American Philosophical Society, volume 1, numbers 1 through 3; signed by CD as Secretary of the Geological Society of London.

14 [1838][1]

To [Richard] OWEN; no location
ALS; $7\frac{1}{4}$ X $4\frac{1}{2}$; 2p.

"I leave you the remaining proofs of yr. descript[ion] of Toxodon and a revise of first part[2]...read quickly over my part. I hope there are no errata left....inform me whether you will want a second revise of your first part....I have made...remarks hap-hazard.... Have you looked at [Alcide Dessalines] D'Orbigny's travels?[3] If not,...you misunderstood...what I mentioned...[so] I have written it [correctly] below...."; P.S. sends duplicate proof for Owen to keep.

1. The Toxodon portion of the first part of the Zoology of the Voyage of the Beagle was published in early 1838. See: "Darwin's Journal," 8; and Freeman, p. 12.
2. See "Toxodon Platensis" in Darwin, Zoology of the Voyage of the Beagle, Pt. I: Fossil Mammalia, by Richard Owen ([1838]-1843), 16-35.
3. Orbigny, Voyage dans l'Amérique Méridionale..., 9v. (Paris: Strasbourg, 1835-47).

15 [1839 January; end. Jan. 1839]

To Cha[rle]s LYELL; no location
ALS; $7\frac{1}{2}$ X $4\frac{1}{2}$; 7p. and add. [Chas. Lyell Esqr/ 16 Hart St.], end. [1a./ Darwin Jan. 1839/ Glen Roy]

Sends Glen Roy paper,[1] which is legible but "ugly from my corrections"; hopes it will not be shortened, as "there is scarcely a sentence, that I have not considered whether I could strike it out, without injuring the...argument"; Lyell may keep paper to read if desired; returns books; last letter of Mr Blackadder [of Glamis] not worth mentioning; made note of information about decaying shells; regrets not having seen "Mr [Charles] Maclaren's capital chapters on alluvium" before writing the appendix, as he upsets CD's "argument of...fixed position of the boulders when drifted, but...confirms...origin of the scratches & grooves."; Maclaren's remarks on boulder positions erroneous, based on "misapprehension, that icebergs drop their cargoes out at sea", which CD's appendix claims is the exception to the rule; ought to have map and will soon have drawing to publish with Glen Roy paper; wishes to discuss "small amount of Alluvial action in Lochaber: occurring since "the sea retired.-- No one point interested me more...."

1. Darwin, "Observations on the Parallel Roads of Glen Roy,..." Phil. Trans. R. Soc., 129 (1839), 39-81. Published version has map and drawing; note on decaying shells is on pp. 63-64.

16 1839 May 23rd

To The Secretary of the American Philos[ophical] Soc[iety],
[Franklin BACHE]; Geological Society of London, Somerset House
Printed L, filled in in ms. (not in CD's hand), S by CD; $11\frac{1}{2}$ X $9\frac{1}{4}$; 1p.

Routine thanks for Proceedings of the American Philosophical Society, volume 1, number 6; signed by CD as Secretary of the Geological Society of London

17 [1839 November 30] Saturday Evening [pmk. NO 30/ 1839]

To Tho[ma]s [Campbell] EYTON; 12 Upper Gower St
ALS; $9\frac{1}{4}$ X $7\frac{1}{2}$; 4p., add. [Thos. Eyton Esqr/ Donnerville House/ Wellington/ Shropshire]

Thanks for agreeing to examine birds mentioned;[1] birds will go to railroad this evening; they are as follows: 388 [and] 707, Tinochorus -----?, habits described in Journal of Researches, p. 110; 630, Synallaxis maluroides, "is it in structure a Certhia?"; 650, Serpophaga albocorunata Gould, a genus allied to Tyrannula; 721, Furnarius cunicularius, habits described as those of Casarita in Journal of Researches, p. 112; 722, Opetiorhynchus vulgaris; 728, Uppucerthia, interesting to dissect this and two previous, as they are altogether unlike European forms; 1037, Pteroptochos albicollis; 1043, Phytotoma rara, "a most curious finch"; 1050, Trochilus gigas, habits in Journal of Researches, p. 331, [Edward] Blyth has notion about humming birds having unique internal structure[2]; 1157, Pteroptochos Tarnii Gray, habits of this and P. albicollis, above, in Journal of Researches, pp. 329 and 352, worthy of close examination; 1309, common North American rice bird [Dolichonyx oryzivorus, the bobolink]; and two birds without tickets, believed to be Opetiorhynci; read on habits before examining; send account of these specimens in month or five weeks; publication date of next number of "Bird Part" of Zoology of the Voyage of the Beagle is uncertain; thanks for offer of Gallinaceous birds for dissection.

1. See Eyton, "Appendix," in Darwin, Zoology of the Voyage of the Beagle, Pt. III: Birds, by John Gould ([1838]-1843), 147-56. For CD's description of the habits of these birds, see Darwin, Journal of Researches [Freeman 4] (1839), pages as indicated above.

2. Blyth, "Outlines of a New Arrangement of Insessorial Birds," Mag. nat. Hist., 2 (1838), 256-68 and 314-19, esp. 258 and 262.

18 [1839] Thursday [?end. 1839]

To [the publishing firm, Henry Colburn]; 12 Upper Gower St
ALS; 7¼ X 4½; 3p., end.? [1839][1]

Capt. [Robert] FitzRoy thinks it desirable and does not object "to my appending an advertisement of the works, connected with the Beagle's Voyage," to the Journal of Researches; ask [Henry] Colburn for approval; thanks for yesterday's note, but CD plans to "append a fly page [giving] ...notice of my works, to be bound up at end or beginning of the volume.-- I believe there are sufficient [notices] now printed...at Mr Smith, Elder..."; how late can CD send these "so as not to delay the binding...."

1. A date of "1839" is written at the head of the first page of the letter in ink similar to that used by CD, but it does not appear to be in CD's hand. Even if added later and not an endorsement by the recipient, this year appears correct, as the letter discusses Darwin's Journal of Researches [Freeman 4] (1839), the only Darwin work published by Colburn.

19 [1839-1842][1] [wmk. 1839]

To [John Maurice] HERBERT; Maer Hall/ Newcastle Stafford
ALS; 7 3/4 X 5; 3p.

Please send note, giving "date &c &c of the event"; Herbert could tell CD nothing when they last met; "We shall remain in the country (at Shrewsbury & here) for some weeks longer."; CD recovering slowly, but not yet well enough for work; "...good wishes & renewed congratulations...."

1. Darwin visited first Maer, then Shrewsbury, for "some weeks" during 1839, 1840, 1841, 1842, and 1844. Given the watermark of this letter, and CD's unwellness during the visit, I am inclined toward the earlier of these years, and have thus eliminated 1844. To fix the exact year, one would have to identify and date Herbert's "event", probably either his marriage or his ordination. "Darwin's Journal," 9-11.

20 [1840 January 6][1] Monday

To [Thomas Campbell] EYTON; 12 Upper Gower St
AL, S by init.; 7 X 4½; 4p.

Ill, headache daily for week; delighted at Eyton's progress,[2] finds results curious; thanks for undertaking task; "I cannot say when the next (& last) number of Birds will appear", perhaps March 1 or two or three months later; appreciates Eyton's offer to produce engravings of specimens, but "I am anxious to spend the government grant in the best way for science, &...I have already given...too much...to the birds & Mammalia"; Eyton must decide with this in mind; Eyton may keep specimens or give them to College of Surgeons; CD has "become a Father...[as of] last Friday week: it is a little Prince"

1. CD's first child was William Erasmus Darwin, born December 27, 1839, a Friday; see "Darwin's Journal," 9.
2. See letter to Eyton dated November 30, 1839, calendared above. The last number of the birds part of Darwin's Zoology of the Voyage of the Beagle ([1838]-1843) was published in March 1841; it included Eyton's appendix, sans engravings. See: Freeman, p. 13; and Eyton, "Appendix," in Darwin, Zoology of the Voyage of the Beagle, Pt. III: Birds, by John Gould ([1838]-1843), 147-56.

21 [1840 February (19?)] Wednesday morn. [end. Feb. 1840; pmk. FE (19?)/ 1840; wmk. 1839]

To Cha[rle]s LYELL; no location
AL, S by init.; 7 3/4 X 5; 6p. and env., add. [Chas. Lyell Esqr/ 16 Hart St/ Bloomsbury], end. [Mr Darwin Feb. 1840/ Coral reefs in open area/ no deeper than 20 fathoms (106)]

Printed, with minor changes: Life and Letters, I, 301. line 14, change "toughish" to "longish". At end of letter is: the following new points will appear in Darwin, Coral Reefs; CD's belief that coral reefs at greater depths than 20 fathoms in open oceans do not exist (see Journal of Researches [Freeman 4], p. 558) contradicts [Christian Gottfried] Ehrenberg's claim of Red Sea coral beds at 25 fathoms; still, CD's argument that there must have been subsidence in large areas scattered with reefs, originally based on point about coral only at shallow depths, stands anyhow, since areas in which every island is low and formed of coral are immense; CD will use modified system of classifying reefs, namely, lagoon islands or atolls, "'encircling reefs'", fringing reefs, and irregular reefs; modification of conclusion (see Journal of Researches [Freeman 4], p. 567) "will chiefly consist in speaking rather less positively & using the words 'alternate areas' more frequently than 'parallel bands'"; will not discuss distribution of organic forms in Pacific (see Journal of Researches [Freeman 4], p. 568); will come on Saturday, if well.

22 [ca. 1840] [end. Nov. 40; wmk. 1838]

To [John] PHILLIPS; no location
ALS; 9 3/4 X 8; 1p. and add. [Prof. Phillips/ St. Mary's Lodge/ York], end. [Darwin/ Nov. 40]

Encloses copy of paper on earthquakes;[1] has grown older and wiser since he wrote it, so sets "less value on theoretical reasoning in geology"; still thinks there is weight in argument respecting "the necessary slow elevation of mountain chains, which have protuberant axis of Plutonic rock"; welcomes Phillips's comments.

1. Perhaps the manuscript of Darwin, "Observations of Proofs of Recent Elevation on the Coast of Chili [*sic*],..." *Proc. geol. Soc.*, *2* (1848): 446-49.

23 [1841 March 9; ?end. 9th March 1841; wmk. 1839]

To [Charles] LYELL; no location
ALS; 7 3/4 X 5; 7p., ?end. [9th March 1841]

Printed, with minor changes and minor omissions: *More Letters*, II, 173-74 (letter 517). p. 173, line 2, add: "Your objection to objection against upheaval, in favour of glaciers (as explaining Glen Roy) about elevation (you will understand what I mean) is quite new to me and seems very sound."

24 [1841 July] 6th Tuesday [end. june 1841; pmk. JY 6/ 1841]

To Charles LYELL; Shrewsbury
ALS; 9 X 7½; 7p. and add. [Charles Lyell Esqr/ 16 Hart St./ Bloomsbury Sqre/ London], end. [Mr Darwin on Coral Islands/ Belizes & Honduras; (107) *Darwin* June 1841/ Coral reefs at Belize/ & Africa covered by mud]

Printed, with minor changes: *More Letters*, II, 193-94 (letter 532). p. 194, line 20, add: discusses neutral tint for uncertain reefs [on CD's plate 3 in *Coral Reefs*]; concerning Red Sea, if [Christian Gottfried] Ehrenberg is correct, then Captain [Sir Fairfax] Moresby's accounts and charts indicate that "the true reefs...are mere fringes to singularly formed land";[1] Ehrenberg, Moresby, and others will all agree if one assumes "that ancient barrier & encircling reefs, formed by subsidence, have... been *uplifted* &...worn down...& are now...fringed by...reefs"; this view too hypothetical for publication by CD; West Indies reefs are also obscure; "the symmetry of reefs seems greatly disturbed every where except in open ocean, or near open ordinary coast-lines". At end of letter is: Bermuda is similar to Bahamas--formed by elevation of ordinary land, with windward edges solidified by growth of *some* coral; health better, but will ail for years, and since "'race is for the strong'", CD "must be content to admire the strides others make in Science.... I shall just crawl on with my S[outh] American work & be as easy as I can."; probably will return on 15 or 16, wants to see Lyell before he departs [to America].

1. See also letter from Darwin to Lyell, February (19?), 1840, above. Ehrenberg's account is probably his *Ueber die Natur und Bildung der Coralleninseln und Corrallenbänke in Rothen Meere* (Berlin: n.p., 1834).

25 [1841] Friday

To [Charles] LYELL; no location
ALS; 8 X 6½; 8p.

Printed, with minor changes and minor omissions: More Letters, II, 148-50 (letter 499). p. 150, line 2, add: "I wish you had in your mind's eye the quantity of solid rock removed on this beach."

26 [1841; wmk. 1839]

To C[harles] LYELL; no location
ALS; 7 3/4 X 5; 7p. and add. [C. Lyell Esqr], end. [Vide p. 268 Agassiz/ on perched rocks of the Alps/ in contradistinction to those of Jura??]

Gives dimensions and elevation of Chiloe Island and relationship to Cordillera; gives composition and geology of Chiloe; "In mentioning blocks on Chiloe put granite first, because I know more certainly that syenite came from Cordillera...."; blocks are strewed on shores of islets and in narrow creeks on coast, where there must have been channels, which after elevation correspond with those of [Jean Louis Rodolphe] Agassiz in valleys of Jura; see forthcoming paper;[1] no fossils with boulder formations, but on them; doubts "perched rocks" on Jura; perched rocks, if on pinnacles, would be "fearful argument for Agassiz's sheet of ice."; Agassiz seems to consider angularity of Jura fragments a difficulty on ordinary meraine or glacier action; make no changes in published explanations, even though "your view is very probable"; "my talk with R[obert] Brown after that with you has knocked me up...."; [crossed out] "I have brought my mind to neglect all negative evidence, especially absence of shells-- Who would have anticipated [Sir Roderick Impey] Murchison's few shells in center of England."

1. Darwin, "On the Distribution of the Erratic Boulders and on the Contemporaneous Unstratified Deposits of South America," Proc. geol. Soc. Lond., 3 (1838-1842), 425-30; and Trans. geol. Soc. Lond., 6 (1842), 415-32.

27 [1841]

To [Charles] LYELL; no location
ALS; 7 3/4 X 5; 6p.

Cannot help regarding subsidence; did Jura have present form when terrestrial animals were embedded?; if not, subsidence may have been small; all areas (forests, mountainsides, and sea-channels) need not have been cold when ice floated; "I don't look at bridge of ice, (or the subsidence, or the absence of shells, for I think I out-Lyell Lyell)" as difficulty; glacier expert [Jean Louis Rodolphe Agassiz] is best evidence, says that Jura erratics "are totally distinct" from those in Alpine valleys; idea of sea of ice carrying rocks in all directions from a small central point is "monstrous"; hopes there are no perched rocks on Jura; did Agassiz find caldron under existing Alpine glaciers?; caldrons are "most inexplicable part of case under every hypothesis"; agrees regarding the arguing of both sides of issue; can give no reasons for supposing Pentlands to be dry shortly before "elevation during ice time."

28 [1842 October 5-7; end. Octr. 7. 1842; pmk. OC 7/ 1842]

To C[harles] LYELL; [Bromley (from pmk.)]
ALS; 8 X 5; 10p. (first leaf missing) and fragment of env., add. [C. Lyell Esqr./ Kinnordy/ Kerriemuir/ N. Britain], end. [(illegible number--PTC)/ Mr Darwin on/ Corals./ Octr. 7. 1842.]

Printed, with minor changes: More Letters, II, 150-51 (letter 500). Before printed portion, add: Suspects some West Indian and Indian corals are same species; "But corals alter their habit so greatly according to where they grow, that the subject, will, I fear, for a long time be involved in great obscurity."; mentions genus Millepora; Crescent Island in Low Archipelago; difficulty of dead reefs not great; such exist, especially in Chagos group, according to Capt. [Sir Fairfax] Moresby; would admit difficulty if reefs as plentiful in tropical seas as vegetation on tropical land; perhaps an increase of small crustaceae in sea or actiniae on shore robs a reef of food, thereby killing it; "...as we see that the presence of reefs is not universal, we ought to expect to find that those same causes, which determine their absence ab origine in some place[s], should have destroyed them in others--"; reefs perish first to leeward side of island; goes to [Geological Society of London] Council meeting tomorrow; Friday morning [October 7] has returned from 2½-hour meeting; discussed candidates for position of curator and librarian; [?William Charles Linnaeus] Martin of Zoological Society judged best; discussed [Edward] Charlesworth's accusations of unfairness against his candidature, considered but rejected a plan to deny his accusations publicly; Charlesworth challenged [Rev. William] Buckland, Lyell, and [Sir Richard] Owen to argue the "whole old question [of the Crag controversy] before the meeting!"; "...it is not the wise who rule the unwise in this world, but the active rule the inactive and verily Charlesworth is...active....";[1] second part of sixth volume of Trans. geol. Soc. Lond. was approved; last Friday, had long talk with [William] Lonsdale, who was cheerful for first time in his life because of [Wollaston Fund] gift, which he will use on coral work--"a noble return" on the gift. p. 151, line 9, change "the sheep" to "two sheep". At end of letter is: excuse length of letter; CD's wife, baby [Mary Eleanor Darwin] "going on fairly well" following birth, son William stronger [?after illness].

1. For more on the topics discussed at this Council meeting, see Horace B. Woodward, The History of the Geological Society of London (London: Geological Society, 1907), 148. On Charlesworth's challenge concerning the Crag question, see Lyell: The Years to 1841, chapter 14, passim.

29 [ca. 1842] Sunday [wmk. 1840]

To [William Hallowes] MILLER; Down (type 1)
ALS; 8 X 5; 4p.

After two-three year interval, is preparing geological notes;[1] some blanks in MS where description of mineral specimens, sent to Miller, should be; has searched specimens returned by Miller and cannot find these; first missing specimen, according to notebook, is "'378 (yellow) a prism of 79½, not yet ascertained'...in which the cells are half filled up horizontally"; since not yet ascertained, CD presumes Miller has it; other missing specimens, "240 & 246 (white)", were deposited with Miller, are "what [Adam] Sedgwick would...call 'beastly rocks'", and "form an entire island, though...a small one"; reply soon; has been ill; "I have left London & bought this place [Down House] & I find the change very agreeable."; those interested in welfare of Geological Society of London should attend special general meeting on December 3.

1. This appears to refer to the manuscript of Darwin, Volcanic Islands (1844); according to "Darwin's Journal", 10, CD began to revise Syms Covington's manuscript for this book on October 14, 1842, which is the reason for the date for this letter as given above.

30 [1842]

To [Charles LYELL]; no location
AL, S by init.; 8 X 6½; 5p.

Printed in full, with minor changes: More Letters, II, 195-197 (letter 533). p. 196, line 31, questionable word is definitely "thickness", and italicize and underline "may".

31 1843 June 25

To [?William Jackson] HOOKER; Down House/ Orpington Kent
L or copy of L; 8 3/4 X 6 3/4; 1p.

Thanks for information; has sent notes on Volcanic Islands,[1] which please return after reading so CD can rewrite and correct them; has been very busy "since I came here [Down House]"; suits him at new place; is determined "to show the people the gift of mankind in regards penmanship of an unusual kind...."

1. Undoubtedly the MS of Darwin, Volcanic Islands (1844).

32 [1843 September] Friday [end. Sept. 1843; wmk. 1841]

To [Charles] LYELL; Down (type 2)
ALS; 7 X 4½ (black border); 8p., end. [(6) Darwin on Kemp/ Sept. 1843/ germinatn of fossil seeds]

Printed, with minor changes: More Letters, II, 244 (letter 577). At beginning of letter is: does not know if Lyell is still in Kinnordy; has not been to London for awhile because of wife's advanced pregnancy;

at British Association meeting [in Cork the preceding August], geological department was poor, but zoological better than usual; has had "sanguine letters from [George Robert] Waterhouse", who was grateful for Lyell's aid; if Waterhouse is hired, he will "enjoy his seven shillings a day from the British Museum, as much as most men would ten times the sum..."; "forlorn" letter from [William] Lonsdale, concerning "very fine series of Touraine corals"; [Edward] Forbes lent Lonsdale his recent Mediterranean species; Lonsdale's only identification as yet with recent species is with "a curious, undescribed Escharina from Dartmouth harbour!" line 15, change "a vivification" to "the revivification". line 22, add: began working a MS in October, has "cut away & shortened at a good rate"; two years ago, thought MS was fit for publication; Lonsdale will describe "corallines from a (mountain limestone?) series from Van Dieman's Land". At end of letter is: greetings to Lyell family.

33 [1843 December 16] Saturday [end. Decr. 1843; pmk. DE 16/ 1843; wmk. 1842]

To C[harles] LYELL; Down (type 1)
ALS; 10 X 8; 3p. and add. [C. Lyell Esqr/ 16 Hart St/ Bloomsbury Square/ London], end. [Darwin 99a/ Tosca or Pampean mud/ not diluvium. Sections/ of it--/ Decr. 1843.], sketches

Has consulted notes, finds "that the proposition that the Tosca was a diluvial mud is monstrous", since it "is distinctly stratified in some parts"; Tosca "not the last deposit"; gives illustrations, with sketches of sections, from Uruguay River and Banda Oriental; "there appears to be an older & newer tosca"; upper tosca, with exceptions, is similar "over wide spaces"; implies a disagreement over this point with [Alcide Dessalines] D'Orbigny, gives example of Rio Negro to illustrate D'Orbigny's error; finds "that the comglomerate of pumice in sandstone in the Patagonian Tertiary is apocryphal."

34 [ca. 1843-1846 or 1855-1861][1] March 26th

To ?; Down (type 2)
ALS; 8 X 5; 1p.

Returns Greenland Catalogue with thanks; is ashamed to have forgotten to return it sooner.

1. Years determined by Down address variant used.

35 [ca. 1843-1846 or 1855-1861][1] June 3d

To [Henry DENNY][2]; Down (type 2)
ALS; $7\frac{1}{4}$ X $4\frac{1}{2}$; 4p.

Thanks for note; corr. is "at perfect liberty to mention Mr. Martial's story";[3] Martial was a ship's surgeon, but worthless and slightly educated; "perhaps, however, in some respects his story is less likely from this cause to have been invented.-- I myself do not think our supposed knowledge of having come from one stock ought to enter into any scientific reasoning"; Eastern and Western Europeans have different species of intestinal worms; cannot now search for specimens, but will do so later if requested; Pediculi perish on wild animals during passage to England, and a slight fever or *broken wrist with no fever* can cause evacuation of intestinal worms, which shows that slight changes in constitution affect parasites.

1. Years determined by Down address variant used.
2. Written on original letter by CD's son was Denny's name.
3. See Darwin, *Descent of Man* (1871), I, 219.

36 [ca. 1843-1846 or 1855-1861][1] Sept 22

To ?; Down (type 2)
ANS; 8 X 5; 1p.

Sends short and interesting addition to be tacked on to end of [?Thomas Henry] Farrer's paper, if corr. prints it.

1. Years determined by Down address variant used.

37 [?1844][1] Jan. 20th

To [Henry DENNY]; Down (type 1)
ALS; $7\frac{1}{4}$ X $4\frac{1}{2}$; 3p.

Would like to help Denny; collected some lice, but part of collection was lost and CD has been prevented from going over his zoological collection by ill health and desire to finish geological works; will go over collection soon and will then save lice for Denny; [George Robert] Waterhouse can help identify CD's ticketed specimens for Denny.

1. CD sorted his collections in mid-1844, after finishing his *Volcanic Islands* (1844); see "Darwin's Journal," 10-11.

38 1844 Aug 29th [end. 29 Aug 44; wmk. 1842]

To [Leonard] HORNER; Down (type 1)
ALS; $7\frac{1}{4}$ X $4\frac{1}{2}$; 7p. and end. [C. Darwin/ 29 Aug 44]

Printed, with minor changes and minor omissions: *More Letters*, II, 115-17 (letter 480). p. 116, line 15, questionable word is definitely "relieved". p. 116, line 25, change "unfilled" to "upfilled". At end of letter is: Emma [Wedgwood Darwin] will give instructions for travel to Down.

39 [1844 September] Sunday [end. Sept. 1844; wmk. 1842]

To C[harles] LYELL; Down (type 1)
ALS; $7\frac{1}{4}$ X $4\frac{1}{2}$; 7p. and fragment of env., add. [C. Lyell Esqr/ 16 Hart St/ Bloomsbury Square/ London], end. [Sept. 1844/ Darwin/ Patagonia/ rising gragually--/ Mastodon/ D'Orbigny/ sudden upthrust of/ Patagonia controverted], sketch

Printed, with minor changes and minor omissions: More Letters, II, 117-20 (letter 481). p. 119, line 22, change "Pampas [debacle?]" to "Pampaean debacle [i.e. Pompeian earthquake]". p. 120, line 17, add: regards to Lyell's wife and had hoped to have seen Horners at Down.

40 [?1845] Jan 26th [wmk. 1844]

To ?; Down (type 1)
ALS; $7\frac{1}{4}$ X $4\frac{1}{2}$; 6p.

Ill; extract, sent by corr., is fullest account CD has seen, refers to island called Pouynipete or Seniavane [i.e. Seniavine or Senyavin]; this is same island as is mentioned in Darwin, Coral Reefs (1842), pp. 127 and 168; describes how he can claim this; description sent by corr. apparently refers to a high island, a significant fact for CD, but "Every one knows how greedily a theorist pounces on a fact, highly favourable to his views," thus CD wished to believe in this fact; nevertheless, CD skeptical of it; writer spoke of granite blocks, but CD thinks island is volcanic; moreover, CD "heard (perhaps...unjustly) very indifferent accounts of Dr. Lloghtsky's moral character; agrees that the case is neither fully established in fact nor fully fabricated; "I have very little doubt that hereafter, the existence of former wide tracts of land, since buried in the ocean by subsidence, will turn out the chief means of the migrations & passage of animals, plants & man, from one part of the world to another", gives examples to demonstrate this; apologies for length of letter, thanks for prompt answer, compliments to "Miss Smith".

41 1845 Feb. 7th [wmk. 1844]

To E[dward] W[illiam] BRAYLEY; Down, Kent
ALS; $7\frac{1}{4}$ X $4\frac{1}{2}$; 2p. and add. [E. W. Brayley Esqr]

Letter of recommendation. Brayley apparently applying for a lectureship in geology; praises Brayley's "remarkable powers in acquiring scientific knowledge of varied kinds, &...your extensive reading."

42 [1845 February 8] Saturday [end. 1845; pmk. FE 8/ 1845; wmk. 1842]

To C[harles] LYELL; Down (type 2)
ALS; 10 X 8; 3p. and add. [C. Lyell Esqr/ 16 Hart St./ Bloomsbury Sqr/ London], end. [Darwin (9) 1845/ Dorbigny on S. Amer/ican shells common] to Europe, Silur-/ian & Cretaceous/ 1st. & 2d. p. of letter]

[Alcide Dessalines] d'Orbigny describes ten Silurian fossils from eastern Bolivian Cordillera as similar to European species, does same for seven Devonian fossils and 23 Carboniferous fossils, although two of the latter, viz. Natica antisinensis [i.e. antisiensis] & Spirifer Roissyi, are not new species; five of the cretaceous fossils, says D'Orbigny, are common to Paris Basin;[1] forgot, when with Lyell on Thursday, to ask Lyell to speak again to [John] Murray about CD's Journal of Researches [Freeman 7 or 8] (1845); since their meeting, saw [Hugh] Cuming about South American fossils and "their range with respect to my Tertiary species"; only series Cuming has not examined and wishes to examine are about 90 shells from Patagonia and Tierra del Fuego; wishes to have this done for several reasons; discusses arrangements to do so; "I fear you will think this so much trouble, that you will wish I had never given you my collection."

1. Letter explicitly refers to Orbigny, Voyage dans l'Amérique Méridionale..., v.3, pt. 3: Geologie (Paris: Strasbourg, 1842), 226, 230 [sic; should be 233], and 239.

43 [1845 July] Saturday [wmk. 1845]

To [Charles] LYELL; Down (type 2)
ALS; 10 X 8; 4p.

Printed, with minor changes and minor omissions: Life and Letters, I, 337-39. p. 338, line 16, change "the first" to "this first". p. 338, last line, add: remembrances to Lyell's wife; CD's wife remains "wearisme". At end of letter is: remembrances to Lyell family at Kinnordy.

44 [1845 August 2] (Saturday) [end. 1845 and in another place Augt. 1. 1845; wmk. 1842]

To [Charles] LYELL; Down (type 2)
ALS; 10 X 8; 9p. and end. [(4) 1845/ C. Darwin/ Criticisms on/ Lyells Travels in/ U.S &c/ (4)] and [Darwin/ Augt. 1. 1845]

Printed: Life and Letters, I, 339-41. At end of letter is: remarks on eight particular passages in Lyell's new book,[1] as follows: v. 1, p. 81, on resemblance of corals, shells, & insects and on analogy of Arctic and Antarctic fauna; v. 1, p. 138, on extinct species of Fulgur and Gnathodon; v. 1, p. 150, on breathing as source of carbonic acid; v. 1, p. 181, on means of water-erosion on seam of carbon; v. 2, p. 37, on Fuegians using a hollowed tree; v. 2, p. 54, on wood or fruits floating on sea; v. 2, p. 65, on buffaloes killed while rushing to drink; v. 2, p. 189, on parallels between present Arctic and Lyell's Carboniferous floras, in terms of extent of distribution; "Might you not...bring more prominently forward the absurdity of arguing from one quarter of the globe, without knowing what was going on in other parts...."; CD's wife and baby [George Howard Darwin] are well; further family details; "P.S. Have you any of my volumes of Lamarck??"

1. Lyell, Travels in North America..., 2v. (London: John Murray, 1845).

45 [1845] Aug. 25th.-- [wmk. 1842]

To [Charles] LYELL; Down (type 2)
AL (incomplete); 10 X 8; 4p.

Printed, with minor changes: Life and Letters, I, 341-42. p. 341, line 3, add: "Please read this before you go" to America on September 4; concerning radiation of snow, CD opposes "the colour-doctrine.-- I find from [?John] Leslie (in [Andrew] Ure[1]), the radiating...power of Lamp-black being called 100, and gold, silver, copper being 12; Writing paper is 98, plumbago 75 and ice is 85. From [William Charles] Wells,[2] it appears, that when swan-down...exposed to open sky falls 16°; grass falls 15°; & snow falls between 12° & 13°: gravel & flag-stone...are inferior to grass, but how much is not said."; Dr. [Patrick] Wilson gives similar data; concludes that snow-covered land "radiates its heat, but little less than the most favourable land." p. 342, line 12, add: multiple and single creations probably discussed in latest Kosmos, since H.[?Friedrich Heinrich Alexander von Humboldt] discussed this with [Joseph Dalton] Hooker "& Humbolt [sic] is a multiple man."; hopes Lyell's next excursion will be to Sicily, to study evidence for and to refute craters of elevation theory. p. 342, line 21, add: will miss visiting Lyells while they are away. p. 342, line 22, add: will send third part of Darwin, Journal of Researches [Freeman 7, pt. 3] (1845) on Monday.

1. This work might be Ure, A Dictionary of Arts, Manufactures, and Mines (London: Longman, Orme, Brown, Green, & Longmans, 1839), although I could not pinpoint the precise passage to which CD refers.
2. CD refers to Wells, An Essay on Dew... (London: Taylor and Hessey, 1814), esp. 43-50. See also Richard Harrison Shryock, "The Strange Case of Wells' Theory of Natural Selection (1813)...," in M. F. Ashley Montagu, ed., Studies and Essays in the History of Science and Learning...in Homage to George Sarton... (New York: Henry Schuman, [1946]), 195-209.

46 [1845] October 8th.-- [pmk. OC 8/ 1845; wmk. 1842]

To Charles LYELL; Shrewsbury
ALS; 9 X 7½; 4p., add. [Charles Lyell Esqr/ Post Office/ Boston/ United States], end. [Mr Darwin/ Queries about negroes]

Printed, with minor changes: Life and Letters, I, 343-45. At beginning of letter is: has not written because has "seen hardly anyone & done little"; "...it has been asserted that on the negroes born in N[orth]. America, the lice are larger & of a blacker colour, than the common species; & that the European lice will not live on negroes."; has heard analogous story about men of Sandwich Islands; asks Lyell to check this and to send specimens of lice from blacks to [Henry] Denny; [Edward] Long's History of Jamaica [London: T. Lowndes, 1774] states that mulattos cross sterile; asks Lyell for comparative information on crosses of "Indians & Europeans & Negroes & Europeans". p. 344, line 14, change "our scientific" to "non-scientific". p. 344, line 17, missing name is [William John] Broderip.

47 [1845-1848][1] Friday

To [?Isaac ANDERSON-HENRY]; Shrewsbury
ALS; $7\frac{1}{4}$ X $4\frac{1}{2}$; 3p.

Is visiting his father; thanks for offer to experiment on hybrids; please record "all facts, such as the number of plants you experimentise on, their names &c &c.--"; "Negative facts (ie failures) are as important to know as successes.--"; will acknowledge source of all results published; sends copy of Darwin, Journal of Researches [Freeman 8] (1845), which "is I hope somewhat improved, from the 1st [edition] that was published."

1. The second edition of the Journal of Researches was published--or printed, at least--in 1845; CD's father died in 1848. These set the endpoints for the date. Anderson-Henry was chosen as correspondent merely because this letter was purchased in a lot with another letter to Anderson-Henry and apparently was glued in a scrapbook along with the other letter at one time.

48 [1846 May 12] Tuesday

To [Richard] OWEN; Down (type 2)
ALS; $7\frac{1}{4}$ X $4\frac{1}{2}$; 2p.

Wishes to see Owen on Thursday morning concerning Mammifers of the Plata; if Owen cannot see CD then, send note to "7 Park St Grosvenor Sqr"; has begun reading Owen on British fossils [i.e. A History of British Fossil Mammals and Birds (London: John Van Voorst, 1846)].

1. Date determined by Sydney Smith, St. Catharine's College, Cambridge University.

49 [1846 August 8] Saturday [end. Augt. 10--1846--; pmk. AU 8/ 1846; wmk. 1842]

To Charles LYELL; Shrewsbury
ALS; 10 X 8; 4p., add. [Charles Lyell Esqr. Junr.--/ Kinnordy/ Kirriemuir/ Scotland], end. [(11)/ C. Darwin/ Augt. 10--1846--/ on his work on/ volcanos/ on hybrids]

First portion printed, with minor changes: More Letters, II, 222-24 (letter 557). At end of this portion is: pleased that Lyell will read Darwin, Volcanic Islands (1844); cost 18 months work, but few have read it; "now [ that Lyell is reading it ] I shall feel whatever little (& little it is) there is confirmatory...will work its effect...."; wishes he could say same for Darwin, South America (1846), but cannot; wanted to discuss with Lyell the foliation of metamorphic schists, the absence of recent conchiferous deposits, and the deposit of tertiary formations during subsidence; has corrected two-thirds of South America and hopes to publish during August; returns to Down on Tuesday; family ill; sends regards; must do proofreading.
Next portion printed, with minor changes: Life and Letters, I, 327, lines 3-7. At end of this portion is: regards to wife.

50 [1846 October 3; end. 1846; pmk. OC 3/ 1846; wmk. 1842]

To C[harles] LYELL; no location
ALS; 10 X 8; 3p. and add. [C. Lyell Esqr/ 11. Harley St/ Cavendish Sqr/ London], end. [(10) Darwin on Ramsays/ paper on Denuda/tion/ 1846--]

Printed in full, with minor changes: More Letters, II, 120-22 (letter 482). p. 121, line 23, change "foundations" to "formation".

51 [1846 October 19] Monday night [end. Oct 20, 1846]

To [Smith, Elder, and Company]; 7. Park St./ Grosvenor Sqr
ALS; 7 X 4½; 3p. and end. [C. Darwin by/ Park St./ Oct 20, 1846]

Has received his copy of Darwin, South America (1846); "coloured Plate [i.e. Plate I]" has "its back to all the letter press; it is almost impossible to refer to it" this way; "a stupid trick"; orders corr. to cut out all plates that are bound and make them front the letterpress; likes looks of volume; has not seen any advertisements yet.

52 [?1846][1] Dec. 21st

To [Andrew Crombie RAMSAY]; Down (type 3)
ALS; 7¼ X 4½; 11p.

Ill for week; thanks for letter; glad Ramsay values parts of book [Darwin, South America (1846)]; concerning "traces of Terraces", a "hobby-horse" of CD, did not see signs of such terraces on recent visit to Snowdonia in North Wales; even recent paper[2] on Scandinavian drift by [Roderick Impey] Murchison errs by "speaking of...successive terraces as the direct effects of so many elevations [rather than as] the indirect effect of an elevation, & the direct effect of the sea's destroying power...."; does not know whether old Tertiary beds of South America were submerged until recent layers set down, but certain they were slowly uplifted, with low parts long submerged; "I think this absence of any considerable recent fossiliferous deposits on both E. & W. coasts, the most remarkable thing I observed" in South America; see page 135 of Darwin, South America (1846); this subject "helps to explain the breaks in Geological chronology & has disabused my mind of a prejudice that durable fossiliferous formations are in most places now accumulating."; thinks "ejected volcanic crystals of glassy feldspar are always broken.--"; found Murchison's thin Silurian lava-streams near Stiper Stones to be injected; for measurements of thin streams, see Darwin, Volcanic Islands (1844), 103 and 109; Ramsay working at interesting site, [?Edward] Forbes gave information on it; where lavas are vesicular and decomposed, has seen "most marvellous transitions into sedimentary beds", partly caused by compression and movement of once-solid lava; impossible to say "where lava ended & tuff began, though neither [were] in the lease metamorphosed."; suspects some metamorphosis; doubts alleged high erosive power of gravel on underlying rocks beneath a sea of any depth; apologies for long letter; directs letter to Charing Cross, since Ramsay probably back from Bala [Wales].

1. Year based on publication date of Darwin, South America (1846).
2. Murchison, "On the Superficial Detritus of Sweden,..." Q. Jl. geol. Soc. Lond., 2 (1846), pt. 1: 349-81.

53 [1846] Monday

To [Leonard] HORNER; Down (type 3)
ALS; 7 X 4½; 4p. (enclosure wanting)

Printed in full, with minor changes: More Letters, II, 174-75 (letter 518). Also printed in Katherine M[urray Horner] Lyell, ed. Memoir of Leonard Horner...Consisting of Letters to His Family and from Some of His Friends, ed. by his daughter..., 2v. (London: Women's Printing Soc., Ltd., 1890), II, 103.

54 [?1846][1] Wednesday

To [(?William) HUTTON][1]; Down (type 2) (black border)
ALS; 7 X 4½; 1p.

Thanks for loan of Horticultural Journal; has read Dr. [William] Herbert's paper[2] "with interest"; will return journal to Athenaeum Club; joins wife in regards to "Mrs. Hutton" and family.

1. CD refers to Jl R. hort. Soc. (see Stauffer, ed., CD's Nat. Selection, index entries under William Herbert); first volume published in 1846, thereby setting lower endpoint for date. Articles by Herbert appeared in volumes 1 (1846) and 2 (1847), but in no later volumes. Corr. seems to be William Hutton, because John Lindley was a close friend and colleague of this Hutton and was also a leading figure in the Horticultural Society of London at this time; moreover, William Hutton was a geologist

with an interest in botany and fossils, which suggests both that CD knew him and that he would be interested in horticultural matters. William Hutton died in 1860, thereby setting upper endpoint for date. He was on the island of Malta from 1846 to 1857 (see DNB, 28, 363), leaving only 1846 and 1857-1860 as possible dates for letter. Black border indicates death in Darwin family; only deaths during these years were CD's mother-in-law in 1846 (see Emma Darwin, II, 89) and CD's son and his eldest sister in 1858 (see "Darwin's Journal," 14 and 14n). Earlier year chosen because style of handwriting and ink used (a lighter brown than usual) match style and ink of 1846 letters better than style and ink of 1858 letters.
2. Probably "Local Habitation and Wants of Plants," Jl R. hort. Soc., 1 (1846): 44-49.

55 [ca. 1846-1855][1] Saturday

To W[illiam] B[enjamin] CARPENTER; Down (type 3)
ALS; $7\frac{1}{4}$ X $4\frac{1}{2}$; 4p.

Sorry to have broken engagement, but was unwell; is "most anxious" to have Carpenter's advice; would make special trip [to London] to see him and to order microscope; Carpenter's note convinced CD to get [a microscope] and "I groan to think over the 3 or 4 months [until delivery of the microscope]"; discusses details of possible meeting times.

1. Years determined by Down address variant used.

56 [ca. 1846-1855][1] Saturday

To ?; Down (type 3)
ALS; 7 X $4\frac{1}{2}$; 3p.

Enclosed probably longer than corr. wished, but "a page in Annals swallows up much M.S."; has marked a page for extraction, "without which [page] my remark w[oul]d be unintelligible"; note "expresses my most honest conviction after careful perusal...", despite its laudatory tone; gives permission to alter but not to shorten; offers to proofread galley, "as my style is often very faulty."

1. Years determined by Down address variant used.

57 [1847 January 20; end. Jany/47; pmk. JA 20/ 1847; wmk. 1846]

To C[harles] LYELL; no location
ALS; $7\frac{1}{4}$ X $4\frac{1}{2}$; 3p. and add. [C. Lyell Esqr/ 11. Harley St/ London], end. [(112)/ Darwin -- Jany/47/ Structure of Gneiss]

To show that CD not "overrash in generalising my conclusion", copies passage from Darwin, South America (1846), page 167, which begins on line 6 and ends on line 14 of that page, concerning cleavage and foliation.

58 [1847 January 24] Sunday

To [Charles] LYELL; Down
ALS; 7¼ X 4½; 5p.

Concerning reefs of Tahiti. Supposes "reef under water" to be dead semi-submerged rock separating living reef from "the islets, which Dr Gould calls the Barrier."; Tahiti less perfectly encircled by reefs than other islands of its group, but considers it encircled because of [James] Cook's chart,[1] which has been verified by the French; see page 152 of Darwin, Coral Reefs (1842); reef is much broken where ships enter; Americans unaware of submerged and probably dead part of reef described in Nautical Magazine for 1836;[2] this is the least perfect part, according to Cook; did not color it in plate without consideration; [Joseph Dalton] Hooker visiting Down, working at paper on coal plants[3] and conversing with CD; Hooker admires [Charles James Fox] Bunbury's papers.

1. Perhaps the chart reproduced as Chart V in R. A. Skelton, ed., The Journals of Captain James Cook on His Voyages of Discovery: Charts & Views Drawn by Cook..., printed for the Hakluyt Society (Cambridge: Cambridge Univ. Press, 1955).
2. W. Forbes, "Description of the Reefs on the North-east Coast of Tahiti,..." Nautical Magazine, 5 (1836): 264.
3. Hooker, "On the Vegetation of the Carboniferous Period,..." Mem. geol. Surv. U. K., 2 (1848): 387-430.

59 [1847 March 7] Sunday [end. March 7. 1847; pmk. MR 8/ 1847]

To C[harles] LYELL; Down (type 3)
ALS; 9 3/4 X 8; 3p. and add. [C. Lyell Esqr/ 11 Harley St/ London], end. [(12) Darwin/ March 7. 1847/ R. Chambers on Paralell/ Roads of Glen Roy]

Thanks for copy of seventh ed. of Principles of Geology... (London: John Murray, 1847) with "much new...to refer to"; wants list of new parts. as promised, since CD too busy to read in toto; [?Charles] Stokes has lent CD volumes 1-30 of Annls. Sci. nat.; has been ill; sorry to have missed Lyell while in London; Robert Chambers gave CD a sketch of [David] Milne[-Home]'s views on Glen Roy,[1] and CD has reread his own paper[2] and is "now, that I have heard what is to be said, not even staggered."[3]; Chambers did not read CD's paper with care and did not look at CD's colored map,[4] so "the new shelf...had not been searched for,..."; was "quite chicken-hearted" at Geological Society of London until Lyell reassured him; Darwin, South America (1846) has had "enormous sale" of 100 copies; CD's father better, but "much changed bodily" in last six months.

1. Milne, "On the Parallel Roads of Lochaber,..." Trans. R. Soc. Edinb., 16 (1849): 395-418. See also: Proc. R. Soc. Edinb., 2 (1844-1850): 124-25 and 132-33; and Edinb. new phil. J., 43 (1847): 339-64.
2. Darwin, "Observations on the Parallel Roads of Glen Roy,..." Phil. Trans. R. Soc., 129 (1839): 39-81.
3. See Paul H. Barrett, "Darwin's 'Gigantic Blunder'," Journal of Geological Education, January 1973: 19-28.
4. Darwin, op. cit. (note 2), Plate I.

60 [1847 June] Wednesday [end. June 1847]

To C[harles] LYELL; Down
ALS; 9 3/4 X 8; 3p. and add. [C. Lyell Esqr], end. [C. Darwin/ June 1847/ On Dr.Morton's/ paper on/ Hybridity/ (13)]

Returns [William] Whewell correspondence, likes Lyell's bold reply;[1] returns [volume 3 of] Silliman's J. containing [Samuel George] Morton's article;[2] glad to have seen latter, but thinks it "a merely tabulated compilation from [Edward] Griffith's Cuvier";[3] Morton's worst fault is failure to consult primary sources; gives examples of this concerning dubious hybrids, refers to [Coenraad Jacob] Temminck; "What a capital Journal Silliman's is; there is always something of interest in it."

1. See I. Todhunter, William Whewell..., 2v. (London: Macmillan and Co., 1876), I, 161.
2. Morton, "Hybridity in Animals, Considered in Reference to the Question of the Unity of the Human Species," 39-50 and 203-12.
3. Georges L. C. F. D. de Cuvier, The Animal Kingdom..., with additional descriptions...by Edward Griffith..., 16 v. (London: Geo. B. Whittaker, 1827-1835).

61 [1847][1] Sept 2d

To [Mrs. M. A. T. WHITBY]; Down (type 3)
ALS; 9 3/4 X 8; 3p.

Information on silkworm provided by Whitby "last year at Southampton"; in moths raised from silkworms kept in captivity, are the wings crippled and is flight impossible; is this especially true in France and Italy; if so, are males and females equally flightless; presumes case similar to domestic ducks; has Whitby tried "two experiments on hereditariness" which CD suggested, viz.: first, whether black-eyebrowed caterpillar produces black or dark-eyed caterpillar young; and second, if fat caterpillars called Frales produce moths and, if so, whether moth offspring are "likewise fat & silkless."; needs results, "for in a work which I intend some few years hence to publish on variation, there will be hardly any facts in the insect world."; are there differences in habits in different caterpillar breeds.[2]

1. CD and Whitby met at the September, 1846, meeting of the British Association, held at Southampton; see Whitby, "On the Cultivation of Silk in England," Rep. Br. Ass. Advmt Sci., 16 (1846), pt. 2: 87-88.
2. See Darwin, Variation under Domestication [Freeman 233] (1868), I, 302-03.

62 [?1847] Sept 7th [wmk. 1847]

To ?; Down (type 3)
ALS; $7\frac{1}{4}$ X $4\frac{1}{2}$; 4p.

Some years ago, took his collection of Mollusca in spirits to [?George Brettingham] Sowerby [?the elder]; more interesting forms, including many cirripedes, were then sent to [Richard] Owen; describes physical features of specimen bottles, asks corr. to look for them; "it is most mortifying...to have lost my own Cirripedia, now that I am at work on them."; will be in London in October, will call on corr. "at the [?Royal] College [?of Surgeons] and look over corr.'s cirripede collections; offers corr. a first-stage Scalpellum larva without striation.

63 [1847 October 4] Monday Morning [end. Oct. 1847; pmk. OC 4/ 1847]

To C[harles] LYELL [sic];[1] no location
ALS; 9 3/4 X 8; 3p. and add. [C. Lyell Esqr--/ 11. Harley St/ London], end. [Darwin (110)/ Oct. 1847/ Glen Roy/ Glacier Theory]

Printed, with minor changes: More Letters, II, 187088 (letter 523). At beginning of letter is: Obliged for barnacles; the one marked "Bergen" is the right one, but its locality is unknown; it is not a Conia; will keep shells, including new one, until review of them is finished; thank husband for note; "what an awful joke...if we had all subscribed for a horrid calf's head!"; will be "grievous" if Coal Saurian proves to be a fish; "I will hope still that [Jean Louis Rodolphe] Agassiz's positive assertions may be disproved by bones, as well as footsteps.--"

1. While letter is addressed to Charles Lyell, the salutation greets his wife, Mary Elizabeth Horner Lyell.

64 [1847 October 11] Monday

To [Charles] LYELL; no location
ALS; $7\frac{1}{4}$ X $4\frac{1}{2}$; 4p. and 1p. enc.

Encloses measurements representing results of comparison of Lochaber and Galashiels Terraces, the latter measured by [William] Kemp; results "wonderful", show similar elevations for terraces in two locations; believes meaaurements of [?Alan] Stevenson and [Robert] Chambers are correct, those of [John] Macculloch are wrong; told Chambers that Lyell and CD both thought ice-lake theory worth considering, Chambers replied that this was dream; [Charles] Maclaren did not insert abstract of [David] Milne[-Home]'s paper[1] into Scotsman and thus will not insert CD's letter;[2] [Robert] Jameson will insert it in Philosophical Journal, but CD has written Jameson "to beg him to destroy it."; will return [Casterordes?] paper with [Hugh] Miller's [?book];[3] Down House full of relatives.

1. "On the Parallel Roads of Lochaber,..." Trans. R. Soc. Edinb., 16 (1849); 395-418; Proc. R. Soc. Edinb., 2 (1844-1850); 124-25 and 132-33; Edinb. new phil. J., 43 (1847): 339-64.
2. Printed at the end of Paul H. Barrett, "Darwin's 'Gigantic Blunder'," Journal of Geological Education, January 1973: 19-28.
3. First Impressions of England and Its People (London: J. Johnstone?, 1847); see More Letters, II, 188.

65 [1847 (?ca. October)][1] Saturday

To [Charles] LYELL; no location
ALS; $7\frac{1}{4}$ X $4\frac{1}{2}$; 4p.

Printed, with minor changes: Life and Letters, I, 328n. At end of letter is: delighted by letter[2] from [Bernhard] Studer to [James David] Forbes showing that layers in gneiss have nothing to do with stratification in Alps; this agrees with Darwin, South America (1846); tell [Leonard] Horner of this, as Horner wished to know what things were in the book; enjoyed Lyell's visit to Down; regards to wife.

1. Lyell visited Down in October, 1847; see Life and Letters, I, 360.
2. "Remarks on the Geological Relations of the Gneiss of the Alps," Edinb. new phil. J., 42 (1846-1847): 186-87.

66 1847 Nov. 18th [pmk. 20NO20/ 1847; wmk. 1846]

To [Henri] MILNE-EDWARDS; Down (type 3)
ALS; 9 3/4 X 8; 3p and add. [A Monsieur/ M. Milne Edwards/ Membre de l'Institut/ et Professeur a l'ecole Centrale des Arts/ Paris]

Has, he believes, the male, the female, and the larvae (in different states) of a "singular Lernaea like animal, which is parasitic on Balanus"; these are identified erroneously as the male of the Balanus and as a new genus of isopodous Crustacean parasitic on this male Balanus by [Harry (not Henry, as is printed with article) D. S.] Goodsir in ["On the Sexes, Organs of Reproduction, and Mode of Development, of the Cirripeds,..."] Edinb. new phil. J., 35 (1843): 88; offers specimens of these to Milne-Edwards, "to whose publications, I have long owed much pleasure & instruction"; could send them through Baillieu the Bookseller.

67 [1847-1848][1] Sunday [wmk. 1847]

To [?Robert] HUTTON; Down (type 3)
ALS; $7\frac{1}{4}$ X $4\frac{1}{2}$; 2p.

Thanks to Hutton and friend for help in obtaining introduction to Lady E[mily Georgiana Bagot] Finch[-Hatton, Countess of Winchilsea and Nottingham], but right after seeing Hutton at Geological Society of London, CD heard from his father that an old friend could provide introduction; regards to Hutton family.

1. Watermark is lower endpoint; death of CD's father in 1848 is upper.

68 [ca. 1847-1849] Wednesday 8th

To [Charles] LYELL; Down (type 3)
Copy of L; 10 X 8; 6p. and end. [Darwin/ Letter on Glen Roy/ Milne's paper]

Printed in full, with minor changes: More Letters, II, 181-87 (letter 522). p. 184, line 14, pluralize "lake". Original of this letter is in Cambridge University Library; see Handlist of Darwin Papers, 13.

69 [1847-1855] Sunday [wmk. 1847][1]

To [George Robert] WATERHOUSE; Down (type 3)
ALS; $7\frac{1}{4}$ X $4\frac{1}{2}$; 3p. and end. [C. Darwin Esqr]

Heard Wednesday at Museum that Waterhouse expected back soon; invites Waterhouse for dinner at Down on Saturday the twelfth, return to London Monday morning; [Charles and Mary] Lyell, [Edward] Forbes, [Andrew Crombie] Ramsay, and R[obert Hermann] Schomburgk are also invited; wants to hear "some news of your foreign trip."

1. Upper endpoint set by type of Down address used.

70 [1847-1865] Wednesday Even/ Thursday mg. [wmk. 1847][1]

To Lady [?Harriet Hotham] LUBBOCK; no location
ALS; 7 X $4\frac{1}{2}$; 1p.

Do not worry about missing volume, covered with brown paper, with no title outside; it will turn up some day and will not be needed soon; has received husband's check and will send receipt with microscope, when complete.

1. Lady Harriet Hothem Lubbock's husband died in 1865; it is unlikely that this letter, with its 1847 watermark, refers to the wife of John Lubbock, first Baron Avebury.

71 [ca. 1848 January][1] Saturday night
+
To [John Edward] GRAY; no location
ALS; 7¼ X 4½; 3p. (enclosure wanting)

Regarding loan of the cirripede collection of the British Museum to CD. Enclosed request for collection in groups (pedunculated and sessile separate) is sent to Gray for approval; will correct if Gray disapproves; getting specimens in these two lots is best arrangement for CD; could divide sessile into two sub-groups if necessary; has not mentioned duration of loan for fear of being hampered.

1. See next letter, below, for date.

72 [1848 February 6] Sunday [end. 8 Feb 1848; wmk. 1847]

To [John Edward] GRAY; Down (type 3)
ALS; 7¼ X 4½; 6p., end. [8 Feb 1848] (enclosure wanting)

Received yesterday Gray's note with "good tidings of the great liberality of the Trustees. Now if I do not make a tolerably good monograph, it will be purely my own fault."; encloses thank-you note for Trustees; is not ready for specimens yet, will not be ready for species part for six weeks; will begin with pedunculated division; will consult Gray on size of first loan; sould appreciate names of as many specimens as possible, although this is troublesome; "Without your assistance I shd break down with the synomony [sic; synonymy]."; has all of Mr. Stutchbury [of Bristol]'s collection, which is partly named after British Museum; thanks for "conduct...most generous & handsome".

73 [1848 June 16] Friday [end. 1848; pmk. JU 17/ 1848; wmk. 1847]

To C[harles] LYELL; Down (type 3)
ALS; 7¼ X 4½; 8p. and env., add. [C. Lyell Esqr/ 11. Harley St/ London], end. [(104) Darwin/ Chambers/ paralell/ roads/ 1848]
Printed, with minor changes: Life and Letters, I, 362-63. At beginning of letter is: tells [Mary Elizabeth Horner] Lyell of relief felt by CD and wife over "wonderful escape" of Miss [?Ann] S[?usan] Horner, which CD heard about since seeing Lyell on Wednesday at Council [of Geological Society of London]; "[Leonard] Horner...had a horror of the sea & now it is...justified." p. 362, line 3, missing name is [William] "Buckland". At end of letter is: "If he [Robert Chambers] be, as I believe, the Author of the Vestiges [of the Natural History of Creation] this book [Ancient Sea Margins...] for poverty of intellect is a literary curiosity.-- I have written all this, as I believe it may save you reading the Book; it is to the best of my Belief, an honest account."; shall be in London before Lyell leaves [for Kinnordy]; wishes to visit Lyell then.

74 [1848 June] 28th [end. June 1848; wmk. 1847]

To [John Edward] GRAY; Down (type 3)
ALS; $7\frac{1}{4}$ X $4\frac{1}{2}$; 4p., end. [June 1848]

Will send [some work by Camillo] Ranzani on "Thursday (tomorrow)"; Ranzani work not much use; wants the Conchotrya and Brisnaeus (Brisneus?) [CD's query]; is working at Lithotrya; also wants [Octomeris?] to do when studying [Catophragmus?], especially if there is a specimen "adhering to its support, so that I could get out the dry animal."; apologies for this trouble; "In truth never will a mountain in labour have brought forth such a mouse as my book on the Cirripedia: it is ridiculous the time each species takes me."

75 [1848 June] Wednesday [wmk. 1847]

To [Charles] LYELL; Down (type 3)
ALS; $7\frac{1}{4}$ X $4\frac{1}{2}$; 3p.

Printed in full, with minor changes: Life and Letters, I, 363-64.

76 [1848] Sept. 1st [pmk. 2 SP 2/ 1848; wmk. 1847]

To [Henri] MILNE-EDWARDS; Down (type 3)
ALS; 9 3/4 X 8; 3p. and add. [A Monsieur/ M. Milne Edwards/ Academie Royl. des Sciences/ Paris]

Milne-Edwards's kindness at Oxford [?B.A.A.S. meeting, June 1847] induces CD to ask favor; describes work on cirripedes, gives history of project, identifies collections at his disposal (Cuming, British Museum, etc.), describes methods used; will describe animals within shells as well as shells themselves; asks for help in obtaining loan of specimens, esp. "a single specimen of some of the species figured in the Voyage of the Astrolabe";[1] especially wants genus Alepas; wants the following: Alepas fasciculatus of [Rene Primevere] Lesson; A. parasita of [Jean Rene Constant] Quoy and [Joseph Paul] Gaimard; A. tubulosa do.; Anatifa elongata do. (especially); A. pelagica [Anatife pelagien] do.; A. sessilis do.; A. tricolor do.; A. spinosa do. (especially) (Pollicipes); A. truncata (especially) (Lithotrya); A. sulcata; has found "a good deal new in the Anatomy"; values Milne-Edwards's work on the Crustacea;[2] presumes Milne-Edwards does not care about parasite on Balanus, about which CD wrote [on November 18, 1847; see above].

1. Jules Sébastien César Dumont D'Urville, Voyage de la Corvette l'Astrolabe, 15v. Zoologie, by Jean René Constant Quoy and Joseph Paul Gaimard, 4v. (Paris: J. Testu, 1830-1832).
2. Alcide Dessalines D'Orbigny, Voyage dans l'Amérique Méridionale..., 9v. Vol. 6, pt. 1: Crustaces, by Henri Milne-Edwards and Hippolyte Lucas (Paris: Strasbourg, 1843).

77 [?late 1848][1] Wednesday

To J[ohn] W[illiam] LUBBOCK; no location (black border)[2]
ALS; 7¼ X 4½; 4p.

Thanks for permission to use schoolroom; Mr. Nash will come next Wednesday; has not received drawing; tell son that CD wants to hear about microscope, will see him for half hour [see letter to Lady Lubbock, (1847-1865) Wednesday evening/ Thursday morning, above]; ill; thanks for invitation to meet Mr. Adams and for paper on meteors.[3]

1. See notes 2 and 3, below, for evidences for this date.
2. CD's father died on November 13, 1848.
3. Lubbock, "On Shooting Stars," Lond. Edinb. Dubl. Phil. Mag., 32 (1848): 81-88 and 170-72, and 35 (1849), 356-57; reprinted in Edinb. new phil. J., 44 (1848), 330-31.

78 [1849 June] Friday [wmk. 1847]

To [Charles] LYELL; The Lodge Malvern (black border)
ALS; 8 3/4 X 7½; 4p.

Printed, with minor changes: Life and Letters, I, 376-77. At end of letter is: delighted that Lyell to write new editions;[1] glad to see Lyell's remarks on extermination and "the striking instance of the tree of [John] Bartram";[2] returns home on 30th; ill, must remain idle to be fully cured by Dr. [James Manby Gully]; has bought horse for riding; will attend [B.A.A.S. meeting] at Birmingham [in September] if well; grieved to hear of Lyell family illnesses; sent copy of Darwin, Manual of Scientific Inquiry [Freeman 97] (1849), to Geological Society of London for Lyell; will return two of Lyell's pamphlets on same subject "sometime"; wanted to hear [Roderick Impey] Murchison on Jura-blocks;[3] regards to [Leonard and Anne Lloyd] Horner; expects large sale for Lyell, A Second Visit to the United States..., 2v. (London: John Murray, 1849).

1. Principles of Geology, 8th ed. (London: John Murray, 1850); and Elements of Geology, 3rd ed. (London: John Murray, 1851).
2. See Lyell, A Second Visit to the United States..., 2v. (London: John Murray, 1849), I, 351.
3. "On the Distribution of the Superficial Detritus of the Alps, as Compared with that of Northern Europe," Q. Jl. geol. Soc. Lond., 6 (1850), 65-69.

79 [1849] July 3d. [wmk. 1848]

To [Charles] LYELL; Down (type 3) (black border)
ALS; 7¼ X 4½; 4p.

Printed: More Letters, II, 225 (letter 559). At beginning of letter is: corrections to Lyell, A Second Visit to the United States..., 2v. (London: John Murray, 1849), as follows: v. 1, p. 349, megatherium only found as far south as 39 degrees by CD and not found at all by [Bartholomew James] Sulivan in southern Patagonia; misplaced or omitted words in second volume concerning diameter of a great equatorial telescope and in first volume concerning oxygen and anthracite; left copy of book by Lyell at Malvern with the Wedgwoods.

80 [1849 September 2] Sunday [wmk. 1847]

To [Charles] LYELL; Down (type 3) (black border)
ALS; 8 3/4 X 7¼; 8p.

First portion printed, with minor changes: More Letters, II, 122-25 (letter 483). p. 124, line 5, pluralize "volume" and change "like" to "have liked".
Next portion printed, with minor changes and minor omissions: Life and Letters, I, 377-78. At beginning of this portion is: regards to Lyell's wife from CD's wife; CD's wife will attend [B.A.A.S. meeting in] Birmingham; two Darwin children ill. p. 377, line 13, add: "But yet I somehow liked him better than Ld Mahon [i.e. Philip Henry Stanhope]." p. 378, line 15, change "evolving" to "evoking". At end of letter is: regards to [Charles James Fox and Frances Joanna] Bunbury; sorry to hear that Lyell's father is weak.

81 [?1849][1] Oct. 25th

To G. RANSOME; Down (type 3) (black border)[1]
ALS; 7 X 4¼; 1p.

Is happy to promote Ransome's project; put down CD on subscription list for one pound sterling for the portrait of the bishop.

1. Down variant address used determines endpoints of 1846 and 1855. During this period, two deaths necessitated use of black border, viz. death of CD's father on November 13, 1848, and death of CD's daughter Anne on April 23, 1851. Dimensions and width of black border on this letter match those letters from October, 1849, but not of October, 1851.

82 [?1849 ca. October][1] Saturday Evening [wmk. 1848]

To [Hugh CUMING][2]; Down (type 3) (black border)
ALS; 7 X 4½; 4p.

Has described and named all Cuming's specimens of Pedunculata; will return them at first meeting of Geological Society of London in early November [November 7]; has from Paris a Lithotrya from the Friendly Islands, suspects it is identical to Cuming's single specimen from the Philippines;[3] will Cuming please lend the latter again for comparison, as well as any new pedunculate cirripedes freshly acquired; must review the genera again and write out generic descriptions of the Pedunculata; will then start sessile cirripedes; "I quite dread the genus Balanus."

1. There are three reasons for choosing this date: 1) CD finished the pedunculated cirripedes and commenced the sessile in April, 1850 (see "Darwin's Journal," 12); 2) the November 7, 1849, meeting of the Geological Society of London; and 3) the black-bordered stationery, in mourning for the death of CD's father in November, 1848.
2. This letter is bound in a letter-book amid other letters to Cuming. See also the next note.
3. CD refers to *Lithotrya truncata*; see Darwin, *Recent Lepadidae*, 366-67ff.

83 [1849 November ?1] Thursday

To [Charles] LYELL; no location (black border)
ALS; 8 3/4 X 7½; 6p., sketch

Printed, with minor changes and minor omissions: *More Letters*, II, 126-28 (letter 485). p. 126, line 9, sketch is missing. At end of letter is: will visit London on Wednesday [November 7; see previous letter, above], wants to see Lyell about Royal Medals [for the awarding of which CD voted on November 16 (*More Letters*, II, 131)].

84 [1849 November 18] Sunday [end. Nov. 1849; pmk. NO 20/ 1849]

To Charles LYELL; Down (black border)
ALS; 8 3/4 X 7½; 6p. and add. [Sir Charles Lyell/ Kinnordy/ Kerriemuir/ N. Britain], end. [C. Darwin Etna dikes./ Nov. 1849], sketches, and drawing

Printed, with minor changes and minor omissions: *More Letters*, II, 128-31 (letter 486). p. 131, line 21, add: hopes Mrs. H[enry (i.e. Katherine Murray Horner)] Lyell is well.

85 [1849] Dec 4th

To [Charles] LYELL; Down. (black border)
ALS; 8 3/4 X 7¼; 4p.

Printed: *Life and Letters*, I, 374-75. p. 375, line 12, after "to me.", add: "though really I think it some little reflection on him,,that he did find other & new points to observe." At end of letter is: is now reading volcanic part [of James Dwight Dana's book[1]] which is excellent and original; in last letter [above], claimed that dikes and lava streams never intersect, but now sees that they do so in Sandwich Islands [Hawaii], without cones; thinks this rare, but believes similar cases exist in Galapagos Islands, examples of which CD saw from a distance; Mt. Etna not like this; Dana believes that great Australian valleys are valleys of denudation, have been formed by "running fresh water", but CD unconvinced on latter point; Dana does not discuss craters of elevation; discusses lack of scoriae in Galapagos, abundance at Etna; will be in London on 19th [of December for meeting of Geological Society of London]; "My boasting has done me a deal of good."

Figure 5. Charles Robert Darwin, aged 40. Lithograph by Thomas Herbert Maguire (1821-1895), executed in 1849. (American Philosophical Society)

1. United States Exploring Expedition, 1838-1842, *United States Exploring Expedition [Wilkes Expedition]. During the Years 1838, 1839, 1840, 1841, 1842, under the Command of Charles Wilkes, U.S.N.*, 19v. Vol. X: *Geology*, by James Dwight Dana (Philadelphia: C. Sherman, 1849).

86 [1849] Dec. 10th [end. 1849/ Decr. 10th]

To [Isaac ANDERSON-HENRY]; Down (type 3)
ALS; 7 X 4½; 3p. and end. [1849/ Decr 10th/ C. Darwin]

Was just thinking of Anderson-Henry; thanks for letter; surprised that Anderson-Henry could make so many experiments on Phloxes and mimuli, given that he was also busy with his "removal"; would be grateful for results on this "most curious & interesting subject"; improved health.

87 [1849] Dec. 25th [end. 25th Decr. 1849/ pmk. DE27/ 1849; wmk. 1846]

To Albany HANCOCK; Down (type 3)
ALS; 9 3/4 X 8; 6p. and add. [Albany Hancock Esqr/ St. Mary's Terrace/ Newcastle/ upon Tyne], end. [25th Decr. 1849/ C. Darwin]

Printed in full, with minor changes: "Letters from C. Darwin, Esq., to A. Hancock, Esq.," *Nat. Hist. Trans. Northumb.*, *8* (1886): 256-58. p. 258, line 3, change "Asthrobalanus (=*Cryptophialus*)" to "Arthrobalanus".

88 [1849 December] Friday Even [wmk. 1846]

To [Charles] LYELL; Down (type 3)
ALS; 9 3/4 x 8; 4p.

Printed, with minor changes: *More Letters*, II, 225-26 (letter 560). At beginning of letter is: overflowing dikes on both sides of volcano is an exceptional case, cause of it "appears connected with liquidity or abdence of much gaseous emissions"; Lyell should read [James Dwight Dana's] whole chapter[1] on Hawaii and the summary on vulcanism in Pacific; other volcanic chapters "have little in them"; discusses Dana's evidence for many currents proceeding from fissures; would contradict [Jean Baptiste Armand Louis Leonce] E[lie] de B[eaumont] because Mt. Etna is scoriae-producing, and there must be cones when there is much scoriae; Dana gives woodcut of denudation crater; discusses Dana's estimates of inclinations of lava-streams; Dana's book would have been more valuable if he had not "compared his results with those of others"; differences in liquidity of lava are immense; other details on Dana's volcanic writings. p. 226, line 7, missing phrase is "*far penetrating* [the] country".

1. In United States Exploring Expedition, 1838-1842, *United States Exploring Expedition...*, 19v. Vol. X: *Geology*, by James Dwight Dana (Philadelphia: C. Sherman, 1849).

89 [?late 1849-early 1850][1] Friday

To [Richard] OWEN; Down (type 3)
ALS; $7\frac{1}{4}$ X $4\frac{1}{2}$; 2p.

Wishes particularly to see a valve of a cirripede in [Frederick] Dixon's collection which is shown in figure 9 of Plate XXVIII [sic: XXVII]; also wishes specimens in figures 3 and 4 of Plate XIV.[2]

1. CD's original request to Owen on this subject is in a letter dated September 10, [1849] (Sir Gavin de Beer, "Further Unpublished Letters of Charles Darwin," Ann. Sci., 14 (1958), 102-03). CD finished working on the pedunculated cirripedes and commenced the sessile cirripedes in April, 1850 ("Darwin's Journal," 12). Owen never provided the specimens; see Darwin, Fossil Lepadidae, 22 and 37-38.
2. See Dixon, The Geology and Fossils...of Sussex (London: Longman, Brown, Green and Longmans, 1850), Plates XIV and XXVII.

90 [1850 January 3] Thursday Evening [pmk. ?JY 4/ 1850; wmk. 1846]

To [Charles] LYELL; Down (type 3)
ALS; 9 3/4 X 8; 4p., add. [Sir C. Lyell/ 11. Harley St/ London], end. [Darwin/ Denudation Craters]

Suggestions and comments on Lyell's paper on craters of denudation, read at meeting of Geological Society of London [on December 19, 1849];[1] paper "will be a thorn in the side of [Jean Baptiste Armand Louis Leonce] E[lie] de B[eaumon]t"; Lyell overlooks CD's case of tuff-strata at Galapagos, viz. beds form narrow streams, hollow from setting of crust, so idea of uplifted horizontal strata absurd; Lyell should state subject at issue more clearly; considers St. Jago, Mauritius, and St. Helena, but not Palma, to be craters of denudation; [James David] Forbes's paper on Italian volcanoes[2] has been referred to CD.

1. "On Craters of Denudation,..." Q. Jl. geol. Soc. Lond., 6 (1850): 207-34.
2. "On the Volcanic Formations of the Alban Hills, near Rome," Proc. R. Soc. Edinb., 2 (1844-1850): 259-61.

91 [?1850 January 24][1] Thursday Evening

To [James Scott BOWERBANK][2]; Down (type 3)
ALS; $7\frac{1}{4}$ X $4\frac{1}{2}$; 1p. and end. [Darwin C.]

Thanks for Balani, which will be of use when doing fossil sessile cirripedes; "I got youn[g] Lubbock [i.e. John William, Baron Avebury] to join your [?Palaeontographical] Society".

1. Lubbock joined the Palaeontographical Society in 1850. Month and day derived by Thaddeus J. Trenn from a letter at the New York Botanical Gardens. See also note 2, below.
2. Of all the persons who lent Balani to Darwin, only Bowerbank was closely connected enough to a society for CD to call it "your Society".

92 [1850 March 8] Friday [wmk. 1847]

To [Charles] LYELL; Down (type 3)
ALS; 7¼ X 4½; 7p.

Printed, with minor changes: More Letters, II, 228-29 (letter 562). At end of letter is: regards to Lyell's wife; [Christian Leopold] Von Buch implies that there are no lower Cretaceous beds in the North, but some of [Johannes Japetus Smith] Steenstrup's cirripedes are marked "Grunsand [i.e. Greensand]" from "[Saliberg?] [?in Scania, the southernmost district of Sweden] Quedlingburg [?in Westphalia]" and CD believes they may be from Greensand, "or at least lower chalk"; are there lower Cretaceous beds in Scania or Denmark?

93 [?1850] May 12th [wmk. 1846]

To [Albany HANCOCK]; Down (type 3)
ALS; 7¼ X 4½; 4p.

Printed in full, with minor changes: "Letters from C. Darwin, Esq., to A. Hancock, Esq.," Nat. Hist. Trans. Northumb., 8 (1886): 259-60.

94 [1850 June 8] Saturday

To [Charles] LYELL; Down (type 3)
ALS[1]; 8 X 5; 3p.

Concerning ripples in the sea-bottom, has only seen them to depth of six to ten feet, but see Darwin, Volcanic Islands (1844), page 134, where deeper observations are attributed to M. Sian;[2] width of ripples related to depth; thanks for Theodore Parker, [?A Letter to the People of the United States Touching the Matter of Slavery (Boston: J. Monroe and Co., 1848)]; glad Lyell approved of paper;[3] "[Jean Louis Rodolphe] Agassiz has sent me his Lake Superior Book [Lake Superior,... (Boston: Gould, Kendall and Lincoln, 1850)],--is not that an immense Honour!"

1. On the back of this letter is a one-page ANS from Emma Wedgwood Darwin to Mary Elizabeth Horner Lyell concerning tickets to see the new hippopotamus in the Zoological Gardens; also mentions that Darwins leave for Malvern on Tuesday [June 11, 1850] for a week's stay; mentions Mary Lyell's sister's health.
2. M. Sian, "On the Action of Waves at Great Depths," Edinb. new phil. J., 31 (1841): 245-46.

3. Probably "On British Fossil Lepadidae," Q. Jl. geol. Soc. Lond., 6 (1850): 439-40. Read June 5, 1850.

95 [?1850][1] Aug 31.

To Lady [Maria] HOOKER; Down (type 3)
ALS; 7 3/4 X 5; 3p.

Sorry that Sir William [Jackson Hooker] is ill; thanks for note and extract; had not heard of [Brian Houghton] Hodgson's "Physicogeographical memoir" [?"On the Physical Geography of the Himalayas," J. Asiat. Soc. Beng., 18 (1849): 761-88]; will send comments directly to Hodgson; delighted that "your son [Joseph Dalton Hooker] is enjoying the grand Sylhet Mountains."

1. On the date of Sir William's illness, see Mea Allan, The Hookers of Kew, 1785-1911 (London: Michael Joseph, 1967), 188. On J. D. Hooker's visit to the Sylhet Mountains in late summer, 1850, see Leonard Huxley, Life and Letters of Sir Joseph Dalton Hooker, 2v. (London: John Murray, 1918), I, 332.

96 [?1850] Sept. 10th [pmk. SP 10/ 185 (sic)]

To J[ames] S[cott] BOWERBANK; Down (type 3)
ALS; 7 3/4 X 4 3/4; 2p. and env., add. [J. S. Bowerbank Esqr/ 3. Highbury Grove/ London], [?end.] [Darwin (illegible word PTC)]

Requests permission [of Palaeontographical Society] to have "four or five woodcuts" [see Darwin, Fossil Lepadidae, 9] drawn and engraved by [James de Carle] Sowerby; wants approval of plan to give species descriptions in both Latin and English, following S[earles Valentine] Wood; "how inconvenient to those who never (as I for one never do) bind their books [that the Palaeontographical Society binds its annual parts into a single volume for each year]."

97 [? 1850 November or December] 10th

To [John William] LUBBOCK, [Baron Avebury]; Down (type 2)
ALS; 7 3/4 X 5; 1p.

Please send reference for paper on the metamorphosis of the Pycnogons, which CD believes Lubbock mentioned earlier, as CD wants to tell C[harles] S[pence] Bate about it; is "much knocked up with Mr. [James de Carle] Sowerby."

98 [1850] Dec. 5th [wmk. 1850]

To [the Ray Society]; Down (type 3)
ALS; 8 X 5; 4p.

Thanks for note; will confine himself [in Recent Lepadidae] to eight plates, with two of them colored; [George Brettingham] Sowerby [the younger] has only drawings, not engravings; please return "skeleton Plates, & the Drawings"; wishes to know when in 1851 will [Recent Lepadidae] appear, so he can schedule work on the eight plates after finishing proofs of [Darwin, Fossil Lepadidae] for Palaeontographical Society; is not "dilatory, though my health allows me to work but for a very short time daily."

99 [1851] July 7th

To [?James Scott BOWERBANK]; Down (type 3) (black border)[1]
ALS; $7\frac{1}{4}$ X $4\frac{1}{2}$; 4p.

Please ask Council of Palaeontographical Society if, to save expense, the Ray Society may use the woodcut figure 1 in Darwin, Fossil Lepadidae, page [9], for use in Darwin, Recent Lepadidae, [facing page 3]; send woodcut to G. Snow [CD's agent in London].

1. CD's daughter, Anne Elizabeth Darwin, died on April 23, 1851.

100 [1851] Dec. 19th

To ?; Down (type 3)
ALS; 5 X 8; 1p.

Thanks for note of 16th; Ray Society has already given CD 22 copies of Darwin, [Recent Lepadidae]; cannot complain; if corr. sees a copy for sale, tell CD its price.

1. Pasted to the back of this letter were four compliment cards, apparently unrelated to CD.

101 [?1852][1] Ap. 16th.

To G[eorge Crawford] HYNDMAN; Down (type 3) (black border)[1]
ALS; 7 X $4\frac{1}{2}$; 1p.

Thanks for sending larvae of Balanus; has seen them before; thinks them fine and useful specimens.

1. Type of Down address variant used sets endpoints of 1846 and 1855. Black border was employed only twice on April 16 during this

period, in 1849 and probably in 1852. Black border on this particular letter is too narrow for 1849, as others of the same period have a much more prominent border. In addition, CD published on Balanidae in 1851, which explains why he should have received an unsolicited gift of Balanus specimens in 1852.

102 [?1852-1853] June 7th [wmk. 1850]

To [?Josephus Augustinus Hubertus de BOSQUET]; Down (type 3)
ALS; 10 X 8; 2p.

What name does Bosquet give for Verruca [prisca], a drawing of the valves of which was included with Bosquet's letter of 7 April; does outline of the plate to which adductor muscle is attached in fixed scutum have the "almost angular outline" depicted, or is it broken; is basal point of the free or movable tergum as round as Bosquet depicts, or is it worn by attrition; presumes Bosquet's volume will soon be published.[1]

1. See Darwin, Fossil Balanidae, 43-44.

103 [1853] Feby. 15th [distinctive blue ink used]

To C[harles] LYELL; Down (type 3)
ALS; 7¼ X 4¼; 3p. and add. [Sir C. Lyell], end. [(98) Darwin/ Dodecendrie Monogynie]

Thanks for [Jean Louis Rodolphe] Agassiz, [?Lake Superior: Its Physical Character, Vegetation, and Animals,... (Boston: Gould, Kendall and Lincoln, 1850)], but returns it, as Agassiz already sent CD a copy;[1] thanks for [unspecified] pamphlets of [?Andrew Leith] Adams, who "appears as heteredox [sic] as myself"; "I have just finished dissecting a curious cirripede [?Alcippe lampas], which is female & has successive cups of males attached to her: I found one with 12 males so fixed to her! These males I suspect are the most negative creatures in the world; they have no mouth, no stomach, no thorax, no limbs, no abdomen, they consist wholly of the male reproductive organs in an envelope."[2]

1. See Elizabeth Cary Agassiz, ed., Louis Agassiz: His Life and Correspondence, 2v. (Boston: Houghton, Mifflin and Company, 1886), II, 469-70.
2. See Darwin, Recent Balanidae, 556 and 562.

104 [1853] March 19th. [distinctive blue ink used]

To [the Ray Society]; Down (type 3)
ALS; 7 3/4 X 5; 7p.

MS of [Darwin, *Recent Balanidae*] will not be ready when expected; thought printing could occur at any time during 1853, not by a specific deadline; "vast delay" between printing and publication of [Darwin, *Recent Lepadidae*] led CD to think that there were no real deadlines; expected to have been finished early in 1853, but work has taken "*far longer*" than expected, and CD has been ill; has not been idle for a single day; 20 plates and corresponding MS are ready, but CD must obtain at least six more plates at his own expense and must dissect for six more weeks; must rest a few weeks before going to press; will not send MS until beginning of August; asks approval for one colored and two half-colored plates to be struck by [George Brettingham] Sowerby [the younger], at only a few shillings above cost of two colored plates; asks permission for a few woodcuts.

105 [1853] March 24th. [distinctive blue ink used; end. March 25,,1853; pmk. MR25/ 53]

To Charles LYELL; Down (type 3)
ALS; 10 X 8; 5p. (one an insert) and env., add. [Sir Charles Lyell/ 11. Harley St./ London], end. [C. Darwin/ March 25, 1853/ Dana as to volcanos/ being safety valves.]

Printed, with minor changes and minor omissions: *More Letters*, II, 133-135 (letter 488). At end of letter is: has not received letter mentioned by Lyell; will be in London for next meeting of Geological Society of London [on April 6].

106 [1853] Ap. 9th. [distinctive blue ink used]

To [Andrew Crombie] RAMSAY; Down (type 3)
ALS; 8 X 5; 5p.

Interested in foliation and cleavage, pleased by Ramsay's remarks on [George Douglas Campbell, eighth] Duke of Argyll's paper;[1] before he publishes his own paper,[2] which was to have been read with Campbell's paper but which CD did not stay to hear, Ramsay should read Darwin, *South America* (1846), 162-68, esp. 167, on CD's theory that foliation can be determined in some cases by planes of deposition, just as foliation sometimes might supervene on cleavage and sometimes might not; Ramsay seems to have found instances of foliation determined by planes of deposition; has "fought many battles viva voce, with Sir C[harles] Lyell...on the subject...."

1. "On the Granitic District of Inverary, Argyllshire," *Q. Jl. geol. Soc. Lond.*, *9* (1853), 360-66.
2. "On the Physical Structure and Succession of Some of the Lower Palaeozoic Rocks of North Wales and Part of Shropshire. With Notes on the Fossils by J. W. Salter," *Q. Jl. geol. Soc. Lond.*, *9* (1853), 161-179. The paper was read on April 20, not April 6. Ramsay did as CD requested; see note on page 172.

107 [1853] June 7th [end. June 1853; pmk. JU 9]

To Charles LYELL; Down (type 3)
ALS; 10 X 8; 3p. and add. [Sir Charles Lyell/ Commissioner to the/ Great Exhibition/ New York/ U. S.], end. [15/ C. Darwin/ June 1853/ Weald Denudation]

Thanks for two pamphlets, one of them "most useful"; attended June 1 meeting of Geological Society of London, describes proceedings; mentions [Peter C.] Sutherland's paper on ice-action,[1] [Joshua] Trimmer's paper;[2] [Roderick Impey] Murchison forwarded catastrophic cause of flints, so CD advanced Lyell's theory of sub-glacial action; Sutherland "most strongly" confirmed Lyell's belief that stones on the beaches in [France and England? the arctic countries?] were angular; [William] Hopkins accepted CD's theory of straight course cut through irregular terrain by highly plastic icebergs; [Robert] Chambers's "interesting" paper on glaciation[3] reproduces [Jean Louis Rodolphe] Agassiz's idea of hemispheric ice-sheet and "treats all Icebergians with the most supercilious contempt."; missed "battle royal" at [annual election meeting] of Royal Society of London, but Murchison and [Francis] Beaufort "gained the day", and [Edward Augustus] Inglefield was elected, exceeding by one the allowed number of admissions; wife is visiting a sister; leaves on July 1 for a month at Isle of Wight,[4] after which CD will "go to press with my weariful cirripedes [Darwin, Recent Balanidae]."

1. "On the Geological and Glacial Phaenomena of the Coasts of Davis' Strait and Baffin's Bay," Q. Jl. geol. Soc. Lond., 9 (1853), 296-312. The portion of the letter that comments on Sutherland is printed faithfully in Life and Letters, I, 329, next to last paragraph.
2. "On the Southern Termination of the Erratic Tertiaries,..." Q. Jl. geol. Soc. Lond., 9 (1853), 282-86; and "On the Origin of the Soils which Cover the Chalk of Kent, Part 3," ibid., 286-96.
3. "On the Glacial Phenomena in Scotland and Some Parts of England," Edinb. new phil. J., 54 (1853), 229-82.
4. The Darwins left instead for Eastbourne, Brighton, and Hastings on July 14; see "Darwin's Journal," 13.

108 [1854] Feb 18th. [pmk. FE19/ 1854]

To Charles LYELL; Down (type 3)
ALS; 10 X 8; 4p., add. [Sir Charles Lyell/ care of H. Murray Esqr/ British Consul/ Santa Cruz/ Teneriffe/ Canary Islands], end. [C. Darwin/ sent to Madeira/ steeper dip of lateral volcc. geoly. than of/ central beds]

Printed in full, with minor changes: Life and Letters, I, 390-92. p. 391, line 1, change "Searle[?]" to "Seale".[1]

1. See Robert F. Seale, The Geognosy of the Island of St. Helena... (London: Ackermann & Co., 1834). Lyell mentions this book in his Elements of Geology, 2nd ed., 2v. (London: John Murray, 1841), II, 227.

109 [1854] Nov. 17th. [wmk. 1853]

To [John Stevens] HENSLOW; Down (type 3)
ALS; 8 X 5; 4p.

Printed in full, with minor changes: Nora Barlow, ed., Darwin and Henslow: The Growth of an Idea (London: John Murray, 1967), 172 (letter 78).

110 [1855] Jany 10th [end. Jany 11, 1855; pmk. JA 10/ 55; wmk. 1849]

To Charles LYELL; Down (type 3)
ALS; 10 X 8; 3p. and add. [Sir Charles Lyell/ 53 Harley St/ London], end. [(19) Ch. Darwin/ foliation of gneiss/ Jany 11,/ 1855.]

Printed, with minor changes and minor omissions: More Letters, II, 204-05 (letter 540). At end of letter is: has found [Daniel] Sharpe's paper ["On the Structure of Mont Blanc and Its Environs," Q. Jl. geol. Soc. Lond., 11 (1855): 11-26]; children are recovering; will take a house in London for four weeks.

111 [1855] Jan 14th [wmk. 1849]

To [Charles] LYELL; Down (type 3)
ALS; 10 X 8; 4p. and sketch

Printed in full, with minor changes: More Letters, II, 205-07 (letter 541). Sketch printed at end of printed letter.

112 [1855 January 21 or 28, or February 4 or 11] Sunday [wmk. 1853]

To [Charles] LYELL; 27 York Place, Baker St[1]
ALS; 8 X 5; 8p., sketch

Printed in full, with minor changes: More Letters, II, 207-09 (letter 542). Sketch printed in middle of page 208. p. 209, line 10, add: "(Do read this P.S.)".

1. The Darwins were at this address from January 18 to February 15; see "Darwin's Journal," 14.

113 [1855] May 8th

To [Charles] LYELL; Down.
ALS; 8 X 5; 4p.

Printed in full, with minor changes: More Letters, II, 153-54 (letter 502).

114 [1855] Oct. 25th

To [Thomas Campbell] EYTON; Down (type 3)
ALS; 7 3/4 X 4 3/4; 3p.

Crayford is too far away for CD to know anything about "Mrs. Shaw"; glad Eyton is thinking of dogs, an "excellent continuation of your capital Pig-Skeleton researches";[1] impressed by Eyton museum [see DNB, XVIII, 107]; family details; recalls hobby of beetle collecting as a youth.

1. Eyton, "Some Osteological Peculiarities in Different Skeletons of the Genus Sus," Proc. zool. Soc. Lond., 5 (1837): 23.

115 [1855] Nov 4th

To [Charles] LYELL; Down (type 2)
ALS; 8 X 5; 4p.

Returns two pamphlets[1] by [John] Bachman; surprised at their poor quality and at Bachman's unsubstantiated assertions, but, as [William Henry] Fitton said of [William] Whewell, "'one must make allowance for him [sic] having sworn to what he believes in'"; "It is most useful to see what is said on all sides" and to read "out-of-the-way pamphlets of this nature"; has living pairs of seven or eight kinds of pigeon which shall be observed and then skeletonized; has begun "to cultivate varieties of plants & make hybrids, so that I have entered on my subject in earnest"; invites Lyells to Down to see pigeons; J[ohn William] Lubbock [first Baron Avebury] sent [Joseph Dalton] Hooker's New Zealand Flora[2] with [Thomas Vernon] Wollaston; please care for these volumes and return them sometime.

1. Most likely these were: Continuation of the Review of 'Nott and Gliddon's Types of Mankind' (Charleston: James, Williams & Gitsinger, 1855); and An Examination of the Characteristics of Genera and Species as Applicable to the Doctrine of the Unity of the Human Race (Charleston: James, Williams & Gitsinger, 1855). See Br. Mus. Cat., IX, 704; see also Cat. scient. Pap., I, 145-46, and VI, 573.
2. The Botany of the Antarctic Voyage of H.M. Discovery-Ships Erebus and Terror in the Years 1839-1843..., Pt. II: Flora Novae-Zelandiae, 2v. (London: Lovell Reeve, 1853-1855).

116 [1855] Dec. 3d.

To [Thomas Campbell] EYTON; Down (type 2)
ALS; 8 X 5; 4p.

Thanks for useful information; is "well in[to] my subject"; has several pigeons in water, plus many alive, and means to get domestic pigeons from all parts of world; delighted that Eyton is at dogs, which will help both CD and "Science"; offers to Eyton the head of a Chinese dog and, if it should die, the carcass of a young, very pure bred German Spitz dog.

117 [1855] Dec 9th

To [Thomas Campbell] EYTON; Down (type 2)
ALS; 8 X 5; 4p.

Cannot find dog's head [see preceding letter, above]; thought Eyton had stopped skeletonizing, until Eyton told CD otherwise at [B.A.A.S.] meeting at Glascow [in September, 1855]; "I took to the nice work [skeletonizing of pigeons], first owing to my wish to see how much the young of Pigeons & Poultry differed from the old, & I have a collection in Brine of nestling Pigeons & chickens."; will not do more than give differences in skeletons of pigeons, poultry, covey birds, and rabbits; thanks for offer to lend pigeon skeletons (including "Almond Tumbler"); will not publish "for some years"; [William] Yarrell has done much bone work, has recently showed CD "a lot of breastbones"; CD's practice of buying curious pigeon carcasses from dealers could be used by Eyton for dogs; "A Mr [William Bernhard] Tegetmeier" will publish on skulls of Fowls,[1] recently showed CD a collection and "gave me a small series."

1. "On the Remarkable Peculiarities in the Skulls of the Feather-crested Variety of the Domestic Fowl, Known as the Polish," Proc. zool. Soc. Lond., 24 (1856), 366-68. Read on November 25, 1856; specimens from Eyton's collection were exhibited at that meeting.

118 1855 Dec. 10

To [George Henry Kendrick THWAITES]; Down (type 2)
ALS; 8 X 5; 6p.

Hopes Thwaites remembers CD from B.A.A.S. meeting at Oxford [1847]; for years, has collected for, and is now preparing, a work on variation of species; wants observations on "any changes in any introduced or feral plants or animals", esp. domesticated pigeons, poultry, ducks, and rabbits; wants all kinds alive, skeletons when dead; is trying especially to get live pigeons from all over world; wants names of pigeon fanciers in Ceylon; wants native names and any remarkable habits of pigeon breeds long kept in Ceylon or imported from anywhere except England; pay a bird skinner ten or fifteen shillings to skin (leaving bones of legs and wings) any old birds of any fancier which happen to die naturally; wants skins of poultry (except silk or black-skinned) or of domestic Ceylonese ducks, if there be any; wants any well-known fancy breed of pigeon (esp. "semi-wild Dove-House Pigeon") if it has been kept in Ceylon for many generations; there are many such in India; apologies for imposition; solicitations over health; "our mutual friend J[oseph Dalton] Hooker" is well.

119 1855 Decr. 20th [wmk. 1855]

Autographs of Members of "Philosophical Club of Royal Society"
DS; approx. 13 X 8; 2p.

Signatures on one side, minutes of the meeting for this date in pencil on the other side, plus title of the document; CD has signed "Mr. Darwin"; all else is in the hands of others.

120 [ca. 1855-1861][1] March 23d

To [?Henry DENNY]; Down (type 2)
ALS; 7 X 4½; 4p.

Thanks for answer to letter of 28 January; sorry corr. could not observe lice from domestic animals from distant lands; interested in aperea, had concluded that aperea was not progenitor of guinea pigs; does not know what corr. means by "stock-dove"; orders would probably be issued if corr. wrote to Council of Zoological Society, but corr. would have to visit the Gardens frequently to see that orders were carried out; still, corr. would get specimens this way; [Philip Lutley] Sclater would be interested in the birds.

1. Type of Down address variant used was employed from 1843 to 1846 and from 1855 to 1861. CD did not know enough about pigeons in the earlier period to have written the highly technical treatment of the stock dove that is in this letter.

121 [?1856] Jan. 2d. [wmk. 1855]

To [John Maurice] HERBERT; Down (type 2)
ALS; 7 3/4 X 5; 4p.

Thanks for book of poetry; "I shall keep to my dying day an unfading remembrance of the many pleasant hours (especially at Barmouth)[1] which we have spent together"; is permanently ill, so cannot visit; other family details.

1. See: Life and Letters, I, 165-66; and "Darwin's Journal," 6.

122 [?1856] Jan. 18th

To [John] PHILLIPS; Down (type 2)
ALS; 8 X 5; 8p.

Recommends reading of the discussion on cleavage and foliation in chapter VI of Darwin, South America (1846); skim page 140, but concluding remarks are on page 162; see also pages 144, 147, 157, 159, and 163, concerning confused cleavages and irregular strikes of foliation at crosses of geological series; cleavage distinct from stratification; "rocks which have been liquified by heat, sometimes have their crystallized materials so arranged, as almost to deserve to be called foliated..."; see also example in CD's description of the Falkland Islands,[1] pages 270-71; existence of grauwacke with clay-slates is perplexing, since clay-slate is apparently formed in deep and tranquil seas; could pressure which causes cleavage and movement along cleavage planes also break up rock, mingle varieties [of rocks] together "like fragments of ice in a glacier," and subsequently re-cement these fragments together; offers copies of the three volumes of Darwin, Geology of the Voyage of the Beagle.[2]

1. "On the Geology of the Falkland Islands," Q. Jl. geol. Soc. Lond., 2 (1846): 267-74.
2. See also John Phillips, "Report on Cleavage and Foliation in Rocks, and on the Theoretical Explanations of these Phaenomena--Part I," Rep. Br. Ass. Advmt Sci., 26 (1856): pt. 1, 369-96.

123 1856 Jan. 23.

To [Walter] ELLIOT; Down (type 2)
ALS; 8 X 5; 4p.

Reminds Elliot of their acquaintance at [the B.A.A.S. meeting of 1855 in] Glascow; wants some items which they discussed then, viz. measurements of differences in proportions of tigers, and remarks on domestic Pigeons "(& Poultry?)" which are extracted from "some work in an Eastern language"; will consult Ayin Akbaree[1] at the India House; is collecting domestic pigeons and poultry, wants Elliot to obtain a bird skinner to provide Indian skins of old, representative specimens of pigeons and poultry long bred in India, esp. tumblers and carriers; gives instructions for preparation of skins; apologies for imposition.

1. Abu al-Fadhl ibn Mubarak, al-Hindi, Ayeen Akbery; or, The Institutes of the Emperor Akber, 2v., trans. from the original Persian by Francis Gladwin (London: J. Sewell, etc., 1800), esp. I, 270. See also Darwin, Variation under Domestication (1868), I, 205 and 205n.

124 [1856] Jan 26th [pmk. JA27/ 1856; wmk. 1855]

To Mrs. [Katherine Murray Horner] LYELL; Down (type 2)
ALS; 7 3/4 X 5; 7p. and env., add. [Mrs Lyell/ 14 Queens Road/ Gloucester Gate/ Regents Park/ London.]

Printed, with minor changes and minor omissions: More Letters, I, 84-85 (letter 42). p. 85, line 11, illegible word is "rumour". At end of letter is: remembrances to husband, Colonel [Henry] Lyell.

125 1856 March 8th

To [George Henry Kendrick THWAITES]; Down (type 2)
ALS; 8 X 5; 4p.

Hopes Thwaites will publish his "facts on variation," esp. "particulars in regard to the species from different elevations, which show different degrees of capacity for cultivation at a new level"; [Joseph Dalton] Hooker has published a similar case with Himalaya rhododendrons;[1] regarding distribution of alpine plants, is there anything new concerning comparison between vegetation at the greater heights in Ceylon and alpine vegetation in the Himalaya, Neilgherries [i.e. Nilgiris, in southern India], or other maountains; do introduced and recently naturalized plants vary much in Ceylon; "The course of my work makes me more & more sceptical on the eternal immutability of species; yet the difficulties on the other theory of common descent seems to me frightfully great. In my work, which I shall not publish for 2 or 3 or perhaps more years; it is my intention to give, as far as I can & that will be very imperfectly, all the arguments & facts on both sides of the case, stating which side seems to me to preponderate."; wants pigeon skins from Ceylon; prefers to work carefully at varieties of a few animals than to compile brief notices on all domestic breeds; "I have now all English breeds of Pigeons alive, & am carefully observing them, making skeletons & crossing them."; wants ducks, rabbits, and poultry; Dr. [Edward Frederick] Kelaert [i.e. Kelaart] will help with poultry; apologies for imposition, "when a beggar once begins to beg he never knows when to stop!"

1. "On the Climate and Vegetation of the Temperate and Cold Regions of East Nepal and the Sikkim Himalaya Mountains," Jl R. hort. Soc., 7 (1852): 69-131. See also Darwin, Origin (1859), 140.

126 [1856] April 21st

To [Charles] LYELL; Down (type 2)
ALS; 6p. @ 8 X 5, 2p. @ 3 1/4 X 8; 8p., 3 sketches

Lyell's case of lava beds passing into vertical columns is most perplexing, but is a "very important & extraordinary fact," and, as Lyell states, is a strong argument against "upheavement" as a cause of an angle of, say, twelve degrees, since lava columns could not have been formed at an inclination and then shifted to the vertical; lava must still have been moving downward when shrinkage produced columns; ask [William] Hopkins on this; thinks Hopkins or [Edward] Forbes has published a sketch of this phenomenon; has seen lava columns; suspects a similar process in glaciers; Lyell's Madeira expedition [in late 1853] was interesting.

127 [1856] May 3d.

To [Charles] LYELL; Down (type 2)
ALS; 8 X 5; 6p.

Printed, with minor changes: *Life and Letters*, II, 67-68. At beginning of letter is: thanks for letter; wishes Lyell had discussed further the lava columns [see previous letter, above]; has kept list of land shells, has made corrections, and is surprised by Lyell's knowledge; will borrow [Oswald] Heer, [?*Ueber die Haus-Ameise Madeira's* (Zürich: n.p., 1852)] while Lyell is abroad; Lyell's cases of transportal "beat all that I have ever heard...& if any body had put such cases hypothetically I shd have laughed at them"; Colymbetes [water beetle] flew on board *Beagle* 45 miles from land, which surprised [Thomas Vernon] Wollaston; Wollaston and others were at Down; "Wollaston strikes me as quite a first-rate man & very nice & pleasant into the bargain. It is really striking (but almost laughable to me) to notice the change in [Joseph Dalton] Hooker's & [Thomas Henry] Huxley's opinions on species during the last few years."

128 [1856] May 10th [end. Ansd. May 15.56; pmk. MY11/ 1856]

To Dr. [Henry Ambrose] OLDFIELD; Down (type 2)
ALS; 8 X 5; 4p. and env., add. [Dr. Oldfield/ 8 Gloucester Gardens/ Gloucester Terrace/ Hyde Park/ London.--], end. [Ansd. May 15.56.]

[William Sandys Wright] Vaux referred CD to Oldfield because Oldfield once remarked that dogs represented in Assyrian drawings were like the Thibetan [i.e. Tibetan] dogs familiar to Oldfield in Nepaul [i.e. Nepal]; as CD is interested in ancient history of domesticated animals, does Oldfield think this resemblance close; describes resemblances and differences suspected; wants data on other breeds of dogs, poultry, fancy pigeons, and rabbits in Nepal; is collecting pigeons from all over world.

129 [1856][1] June 3d.

To [Samuel Pickworth WOODWARD]; Down (type 2)
ALS; 8 X 5; 2p.

Has just finished [Woodward's] book,[2] has derived "much solid instruction & interest" from it; has written down some questions which he will ask in person when in London in about a fortnight.

1. See *Life and Letters*, II, 73.
2. *A Manual of the Mollusca; or, Rudimentary Treatise of Recent and Fossil Shells* (London: J. Weale, 1851-1856). Published in three parts, in 1851, 1853, and 1856; issued thereafter as a single volume.

130 [?1856] June 4th

To [William Darwin] FOX; Down (type 2)
ALS; 8 X 5; 2p.[1]

Thanks for "a Dorking Cock, more like an ostrich than a simple fowl".

1. On same sheet is a two-page letter from Mrs. Fox to her husband.

131 [1856 June] 16th

To [Charles] LYELL; Down
ALS; 8 X 5; 4p.

Printed, with minor changes and minor omissions: Life and Letters, II, 72. At end of letter is: "When you go abroad you are to lend me [ ?Philip Barker] Webb. & [Oswald] Heer, & can you add [Matthew Fontaine] Maury [the elder] ocean chart; [Samuel Pickworth] Woodward had it some time ago.

132 [1856] Jun 25th. [end. 25 June. 1856]

To [Charles] LYELL; Down (type 2)
Partially ALS and partially LS; 13 X 8¼ 5p. and env., end. [(119)/ C. Darwin/ 25 June. 1856/ On reasons for doubting the/ Atlantis/ & continental extension theory/ in the Recent Period.]

Printed in full, with minor changes: Life and Letters, II, 74-78. p. 76, line 4, questionable word is "Silla". p. 76, line 23, add "are" after "such". p. 77, line 4, change "Lowe" to "Low". p. 77, line 12, pluralize "formation".

133 [1856] July 5th [end. July 5, 1856]

To [Charles] LYELL; Down (type 2)
Partially ALS and partially LS; 13 X 8¼; 3p. and fragment of env., end. [C. Darwin/ July 5, 1856/ Continental Extension./ Volcanos whether in/ areas of elevation or/ of subsidence.]

First portion printed, with minor changes and minor omissions: More Letters, II, 135-37 (letter 489). p. 135, line 12, questionable word is "extensions". p. 136, line 31, add: local volcanic subsidence caused by shrinking of great volcanic piles is supported by frequent coincidence of volcanic tertiary streams and lakes or fresh water beds. At end of this portion is: CD's vague ideas on this subject are worthless; CD's ideas are not dogmatic.
Next portion printed, with minor changes: Life and Letters, II, 71. At end of this portion is: shall not attempt to write a history of the subject, but did mention Lyell's Principles [of Geology..., 3v. (London: John Murray, 1830-1833)]; will want Lyell to look over what CD will write about the Principles.

134 [1856] July 8th [end. July 8, 1856.]

**To Charles LYELL; Down (type 2)**
**ALS; 8 X 5; 8p. and env., add. [Sir Charles Lyell/ 53. Harley St/ London.], end. [C. Darwin/ continental extensions/ July 8. 1856.]**

**Printed in full, with minor changes: More Letters, I, 427-28 (letter 327). Small portion also printed in Life and Letters, II, 78.**

135 [?1856] Augt 21st

To [Thomas Campbell] EYTON; Down (type 3)
ALS; 8 X 5; 4p.

Has been reading Eyton on pigs ["Some Osteological Peculiarities in Different Skeletons of the Genus Sus", Proc. zool. Soc. Lond., [5 (1837): 23]; are offspring of cross between African and common pigs fertile; if Eyton does not know answer, ask Lord [Rowland] Hill; were Hill's African pigs domesticated; from where in Africa do they come; has pigeon skeletons for "every breed alive", but has not compared them yet; will need Eyton's help when he does; is compiling "Book on Variation" [i.e. Darwin, "Natural Selection"], but finds it slow work; family details.

136 [?1856 August] 27th

To [Thomas Campbell] EYTON; Down (type 2)
ALS; 8 X 5; 2p.

Excuse this additional note; thinks Eyton has studied Herefordshire cattle;[1] believes there are two strains of this breed that can be distinguished by color on face; are there other distinguishing characters besides color?

1. Eyton, The Herd Book of Hereford Cattle, 2v. (London: Longman and Co., 1846-1853).

137 [?1856] Aug. 31st

To [Thomas Campbell] EYTON; Down (type 2)
ALS; 8 X 5; 10p.

Thanks for note and promise of more information on pigs; is collecting pig jaws to test [Johann Matthäus] Bechstein's assertion that number of incisors varies greatly in domestic pigs; can Eyton confirm this; approves of illustrations in Eyton's "Stud Book" [i.e. The Herd Book of Hereford Cattle; see previous letter, above], returns plates with thanks; after writing, found source of his ideas, viz. Q. Rev., 1849, page 392, mentioning split in Herefordshire breed;[1] would like to mention this in [Darwin, "Natural Selection"] as a rare instance of the documentation of the origin of "even a sub-breed of a sub-breed";

will also refer to Eyton; comments on Eyton's birds; will quote Eyton's case of geese;[2] did Eyton breed the "grandchildren geese"; has been inquiring in India on same subject; one of most troublesome problems is distribution mechanism for species on distant islands; has tried resistance of seeds in sea water; check to see if dirt sticks to feet of birds;[3] check possibility of seed dispersion through pellets thrown up by owls or hawks; send contents of stomachs of dace and other white fish eaten by birds, so that CD can sow seeds contained therein; if Eyton collects cat skeletons, C[harles] Lyell has an odd Persian carcass and CD knows of another odd specimen; "I have put your words, that you like hearing from old naturalist friends, to a severe test."

1. CD here refers to an anonymous review of Henry Stephens, The Book of the Farm..., 2nd ed., 2v. (Edinburgh and London: W. Blackwood and Sons, 1849), to be found in Q. Rev., 84 (1848-1849): 389-424.
2. Eyton, "Remarks on the Skeletons of the Common Tame Goose, the Chinese Goose, and the Hybrid between the Two," Mag. nat. Hist., 4 (1840): 90-92.
3. See Darwin, Origin (1859), 358-63ff; and Peckham, Variorum Origin, 575-79.

138 [1856][1] Sept. 9th

To [Josephus Augustinus Hubertus de BOSQUET, of Maestricht]; Down (type 3)
Portion of AL; 7 3/4 X 5; 4p., and portion of ALS; 7 3/4 X 5; 2p.

Portion of AL is: thanks for letter; CD "astonished & delighted at your discovery of a Cretacean chthamalus", a curious, publishable discovery; usually does not believe in negative evidence, but deviated with sessile cirripedes (see Darwin, Fossil Lepadidae, 5), and now Bosquet's discovery proves CD wrong to deviate; cautions Bosquet against "ever presuming to say when a new group first...appeared"; on structure of recent chthamalus, see Darwin, Recent Balanidae, 39; see also page 172 on non-existence of extinct Chthamalinae; drawings sent by Bosquet are beautiful; has been "so hard at work for two years at other subjects that cirripedes are gone rather out of my head, which could never boast of a good memory"; compliments Bosquet's ability; Lithotrya specimen must exhibit serrations, important for burrowing, on upper scales of peduncle (see Darwin, Recent Balanidae [sic; Recent Lepadidae] plate VIII, figure 3d); thus tergal margin of scutum in Bosquet's figure 1 seems too simple for Lithotrya; thinks Bosquet's figure 2 is a carina. Portion of ALS is: skins could be sent through a bookseller; wants "one of your good Carriers (old Cock bird) skinned"; wants estimate of speed at which carriers fly long distances (e.g. 200-400 miles); hopes specimens arrive safely.

1. See More Letters, I, 97 (letter 51); also, CD stopped working on cirripedes in 1854, so he would be two years into another subject, as he claimed in this letter, in 1856.

139 [1856][1] Oct. 5th

To /Thomas Campbell_7 EYTON; Down (type 2)
ALS; 8 X 5; 4p.

Does Eyton want a skin, with skull and limbs, of the standard type of country dog in West Africa, sent by Dr. /William Freeman_7 Daniell, who lives in Sierra Leone; if so, tell CD if skull has any peculiarities; after writing the "long troublesome letter" /see letter of August 31, 1856, above_7, CD decided to experiment on hawk pellets in Zoological Gardens; others are observing partridge feet; recently found eleven grains of earth on one bird; awaits word both on /Johann Batthaus_7 Bechstein's claim about incisors of pigs and on origin of Lord /Rowland_7 Hill's crossed African pig /see letters of August 21 and 31, 1856, above_7; supposes Eyton discontinued cross of geese /see letter of August 31, 1856, above_7; is using Eyton's papers.

1. CD's experiments on hawks in the Zoological Gardens were conducted in October, 1856; see Life and Letters, II, 86.

140 /1856_7 Nov. 10th

To /Charles_7 LYELL; Down (type 2) (black border)[1]
ALS; 4p. @ 7¼ X 4½, 1p. @ 8 X 5; 5p.

First portion printed, with minor changes: Life and Letters, II, 85. At beginning of this portion is: is answering Lady /Mary Elizabeth Horner_7 Lyell's note to Emma /Wedgwood Darwin, CD's wife_7, since writing is an exertion for the latter;[2] sorry that Mrs. /Leonard_7 Horner /nee Lloyd_7 is ill; had planned on visiting Lyell in London, but shall not do so until January, owing to wife's condition; last week /November 6_7, CD's aunt, /Sarah Elizabeth_7 Wedgwood, died at Down; supposes Lyell's Madeira paper is ready.
Second portion printed; with minor changes: More Letters, I, 97 (letter 51).

1. See Emma Darwin, II, 176-77, for death of Sarah Wedgwood.
2. See ibid., 178, for CD's wife's confinement.

141 /?1856 December 8_7 Monday

To Lady /Harriet Hotham_7 LUBBOCK; Down.
ALS; 8 X 5; 3p.

Thanks for offer of help, but CD's wife is "going on perfectly well", and as Etty /i.e. CD's daughter Henrietta_7 is ill and Miss Thorley /governess_7 is gone, CD's sons should stay at home; "We have now half-a-dozen Boys".[1]

1. This is obviously a roundabout way of announcing the birth of the sixth son to CD's neighbor; December 8, 1856, was the first Monday after the sixth son's birth. See Emma Darwin, II, 178.

142 [1856][1] Dec. 23d

To [Thomas William St. Clari DAVIDSON]; Down (type 2)
ALS; 8 X 5; 4p.

Wants facts showing that "a variable species is or is not equally variable at all times & places"; thinks [Davidson's] "profound knowledge of Brachiopoda" should provide examples of such facts; has discussed this subject with the late E[dward] Forbes and with [Samuel Pickworth] Woodward.

1. Date for this letter determined by Sydney Smith, St. Catharine's College, Cambridge University.

143 [ca. 1856-1860][1] June 8th

To [Edgar Leopold LAYARD]; Down (type 2)
ALS;[2] 8 X 5; 6p.

Thanks for letter; Layard's Madagascar expedition should yield odd domestic animals; are hybrid cats fertile;[3] wants confirmation of [Martin Heinrich Carl] Licktenstein's [sic; Lichtenstein's] assertion[4] that the domestic dog, similar to C[anis] mesomelas, which is kept by the natives is sometimes crossed profitably with wild species; do promiscuously-crossed mongrel dogs tend toward an ideal type; did Mr. Fry's feral pigeons from Ascension have black bars on wing and white rumps, or were they checkered like common dovecot pigeon; did Fry ever see a North African greyhound with short, curly tail, as pictured by ancients, but which CD doubts;[5] wants specimens of pigeons, ducks, and poultry with "very slight differences"; such differences interest CD "greatly"; doubts that a general synopsis of seafowl exists, but shall check in London.

1. Layard left for South Africa in 1855; CD started using the Bromley address in late 1855, and 1856 was the first June in which he did so. CD published Variation under Domestication, containing material received most certainly in reply to this letter, in 1868; CD stopped using the written Bromley address in about April, 1861, meaning that June, 1860, was the last June in which this address was used.
2. Accompanying this letter are two scraps, one with "Charles Darwin/ Down, Kent/ March 19th, 1873.--" in CD's hand, the other with two or three illegible words, not in CD's hand.
3. See Darwin, Variation under Domestication (1868), I, 44.
4. Lichtenstein, Travels in Southern Africa..., trans. from German by Anne Plumptre, 2v. (London: 1812-1815), II, 272. See also Darwin, op. cit., I, 25.
5. See Darwin, op. cit., I, 17-18, 44, 185f, and 238n.

144 [?ca. 1857][1] Jany. 4th.

To ?; Down (type 2)
ALS; 8 X 5; 2p.

Thanks for note; information corr. will obtain from [Marie Jean Pierre] Flourens will be valuable;[2] has sent Darwin, Journal of Researches [?(1852)] through Williams and Norgate [booksellers]; work on variation will not be published for some years; "I have much to observe, & am keeping for this purpose all the varieties of Pigeons, Poultry, Ducks &c."

1. Date determined by Sydney Smith, St. Catharine's College, Cambridge University.
2. See, perhaps, Stauffer, ed., CD's Nat. Selection, 456 and 456n.

145 [1857] Feb. 11 [end. Feb. 13, 1857]

To [Charles] LYELL; Down (type 2)
ALS; 8 X 5; 7p. and fragment of env., end. [C. Darwin/ On instruction for Explo-/ring ship Novara./ Feb. 13. 1857]

Printed, with minor changes: Life and Letters, II, 93-94. p. 94, lines 3/4, change "Cocos so near" to "[Cocos Is(land) mass?]". p. 94, line 11, change "one" to "n[orth]." p. 94, line 22, pluralize "Island" and change "has" to "have". At end of letter is: "I have just had Helix Pomatia quite alive & hearty after 20 days under sea-water; & this same individual about six-weeks ago had a [salt-water] bath of 7 days.[1] P.S. I have really nothing to suggest to Mr. [David] Forbes.[2] I am delighted to hear about the Coal Plant & Purbeck Fossils.[3]"

1. See Darwin, Origin (1859), 397. 2. Forbes invested in a South American mining company and toured South America from 1857 to 1860 in the company's behalf looking for nickel; Lyell probably asked CD for any advice he might have had for Forbes.
3. See Lyell, Principles of Geology..., 2v., 11th ed. (New York: D. Appleton and Company, 1873), I, 159n.

146 [?1857] June 9th

To [Thomas Campbell] EYTON; Down (type 2)
ALS; 8 X 5; 3p.

Thanks for sheets that complete Eyton's catalogue; is astounded at Eyton's "superb collection"; fears Eyton was unable to check fertility of Lord [Rowland] Hill's African pigs [see letters of August 21 and October 5, 1856, above]; how goes the work on skeletons of dogs; presumes Eyton does not want skin of West African domestic dog [see letter of October 5, 1856, above]; if Eyton breeds horses, CD wants observations from him on coloring of colts; do convolutions in the trachea of males of a single species of bird ever vary much?

147 [?1857 June][1] 26th

To [Thomas Campbell] EYTON; Moor Park, Farnham, Surrey
ALS; 8 X 5; 4p.

Thanks for letter and for reference on hybrid; line on growth of bones of birds is new and interesting; ill, but makes "steady progress in my Book on Variation of Species & on domestic varieties"; watch in Ireland or elsewhere for horses or ponies with bars on legs, as with zebra, or on shoulder and along back, as with ass.[2]

1. The only two years in which CD both worked on his big book (i.e. Stauffer, ed., CD's Nat. Selection) and visited Moor Park on the twenty-sixth of the month were 1857 and 1858; see "Darwin's Journal," 14. I am guessing at June, 1857, because of the earlier letter to Eyton on June 9, 1857 (above), and because I believe this is the best candidate date for Eyton to have told CD about the growth of bones of birds.
2. See Darwin, Origin (1859), 163-67.

148 [ca. 1857][1] 22nd

To [Thomas Campbell] EYTON; Down (type 2)
ALS; 8 X 5; 1p.

Will send the West African dog skin in the evening; [William Freeman] Daniell assured CD that "it was a very characteristic specimen of the pure native dog of Sierra Leone".

1. See previous reference to this skin in CD to Eyton, June 9, [?1857], above.

149 [?1857] Wednesday

To [J. Brodie] INNES; no location
ALS; 4 3/4 X 8; 1p. (enclosure wanting)

Read the enclosed,[1] pages 38 to 91; "[Karl Theodor Ernst] Von Siebold is about the most careful & profound naturalist in Europe."; found in the book facts such as those requested from Innes regarding bees.

1. Siebold, "True Parthogenesis in the Honey-Bee," in On a True Parthenogenesis in Moths and Bees; A Contribution to the History of Reproduction in Animals, tr. by William S. Dallas (London: J. Van Voorst, 1857), 38-91.

150 [1858] Feby. 7th

To [George Henry Kendrick THWAITES]; Down (type 2)
ALS; 8 X 5; 4p.

Thanks for letter of December 28, with information on resistance of plants to Ceylonese climate and on acclimatization of plants to differing elevations; will quote Thwaites about several species having both alpine and lowland forms and about some having and some not having intermediate varieties;[1] is it the lowland forms that have much smaller and more numerous flowers with longer, narrower, and less coriaceous leaves; disbelieves, as does [Joseph Dalton] Hooker, the idea that alpine forms have the character of "hariness [?i.e. hairiness]", so is glad that Thwaites does not mention this trait; "I was lately struck by a remark in U[nited] States Naturalist, namely that introduced or naturalized plants at first overrun the whole country, & then in some degree diminish in numbers.... I can see some likely causes of error in the...remark, & yet the fact in itself seems probable.";[2] suggests the cultivation of alpine forms in the low country so as to observe "changes in successive generations."

1. See Darwin, Origin (1959), 140. 2, Ibid., 64-65.

151 [1858] Ap. 26th [end. April 25th/ 1858]

To [Charles] LYELL; Moor Park, Farnham/ Surrey
ALS; 8 X 5; 4p., end. [April 25th/ 1858]

Printed in full, with minor changes: Life and Letters, II, 112-13.

152 [1858 June] 18th

To [Charles] LYELL; Down (type 2)
ALS; 8 X 5; 4p.

Printed in full, with minor changes: Life and Letters, II, 116-17.

153 [1858 June 25] Friday [end. received/ 29 June 1858/ 59; wmk. 1855]

To [Charles] LYELL; Down (type 2)
ALS; 8 X 5; 8p., end. [received/ 29 June 1858/ 59]

Printed, with minor changes: Life and Letters, II, 117-18. p. 117, line 5, add: "I shd. not have sent off your letter without further reflexion, for I am at present quite upset, but write now to get subject for time out of mind. But I confess it never did occur to me, as it ought, that [Alfred Russel] Wallace could have made any use of your letter." p. 117, line 12, underline "extremely" once and "now" twice. p. 117, line 21, add: "I do not in least believe that that [sic] he originated his views from anything which I wrote to him." p. 118, line 14, add: CD's baby has scarlet fever; Etty [i.e. Henrietta Emma Darwin Litchfield] is weak but recovering.

154 [1858 June] 26th [wmk. 1855]

To [Charles] LYELL; Down.
ALS; 8 X 5; 3p.

Printed, with minor changes: *Life and Letters*, II, 118-19. At end of letter is: Emma [Wedgwood Darwin] and CD thank Lady [Mary Elizabeth Horner] L[yell] for note; Etty [see previous letter, above] is weak; baby is feverish; three children in Down have died of scarlet fever.

155 [1858] July 18th [end. July 18 1858; pmk. JY 18/ 58]

To Charles LYELL; King's Head Hotel/ Sandown/ Isle of Wight (black border)[1]
ALS; $7\frac{1}{4}$ X $4\frac{1}{2}$; 4p. and env., add. [Sir Charles Lyell/ 53 Harley St/ London/ W.], end. [C. Darwin/ July 18 1858/ My Etna paper & theory/ of craters of Elevatn./ C.D.'s work on Species at/ Linn. Socy.]

First portion printed in full, with minor changes: *More Letters*, II, 137 (letter 490).
Next portion printed in full, with minor changes: *Life and Letters*, II, 129-30.

1. CD's son, Charles Waring Darwin, died on June 28, 1858; see "Darwin's Journal," 14.

156 [1858] Aug. 4th

To [Thomas Campbell] EYTON; Norfolk House/ Shanklin/ I. of Wight (black border)
ALS; 7 X $4\frac{1}{2}$; 4p.

Thanks for note and for reference; glad to hear of Eyton's two publications; thinks that on bird skeletons will be "laborious";[1] just before leaving Down, CD arranged skeletons of pigeons preparatory to comparison; may consult Eyton on this; is "drawing up a long abstract [later to become Darwin, *Origin* (1859)] on my notions about Species & Varieties, to be read in parts before Linnean Soc[iet]y"; abstract will be published "late in the autumn"; "My bigger Book [i.e. Darwin, *Natural Selection*] will not be out for some two or three years."

1. See Eyton, *A Catalogue of the Skeletons of Birds in His Possession* (London: n.p., 1858).

157 [1858] Sept. 1 [wmk. 1858]

To [John] PHILLIPS; Down (type 2) (black border)
ALS; 7 X $4\frac{1}{2}$; 3p.

Has just heard that he (CD) is advertised for presidency of Zoological Section of the B.A.A.S.; must decline position for reasons of health; also, cannot attend [B.A.A.S. meeting] at Leeds [on September 22]; lost no time in informing Phillips of "this mistake".

158 [1858] Oct. 4th [wmk. 1858]

To [Thomas Campbell] EYTON; Down (type 2) (black border)
ALS; 7 X $4\frac{1}{2}$; 7p.

Thanks for note; sorry to miss [B.A.A.S. meeting at] Leeds [on September 22]; Eyton's skeleton collection is "splendid"; sees that Eyton is publishing on oysters;[1] will keep Eyton's letter with list of skeletons for future reference; is done with domestic pigeon skeletons and with a monograph on their history, variation, etc., totaling four or five pages, which CD would like to send to Eyton for criticism, along with the few bones which show any diversity; must learn names of some bones from [Hugh] Falconer; on advice of [Charles] Lyell and [Joseph Dalton] Hooker, is preparing abstract of conclusions [Darwin, *Origin* (1859)] "to be published as small book or read before Linn[ean] Society, & this will for some months stop my regular work. The work is too great for me, but if I live I will finish it: indeed three-fourths is done."; what colors of sire and dam will throw a dun colored horse; what is color, at birth, of colt which will turn into dun; did Eyton ever see an ass with double shoulder stripes on both shoulders; Col[onel] Ham[ilton Charles] Smith has heard of such.[2]

1. Eyton, *A History of the Oyster and the Oyster Fisheries* (London: J. Van Voorst, 1858).
2. See: Darwin, *Origin* (1859), 163-67; and *idem*, *Variation under Domestication* (1868), I, 55-64.

159 [?1858][1] Oct. 11 [wmk. 1857]

To [Thomas Campbell] EYTON; Down (type 2) (black border)
ALS; $7\frac{1}{4}$ X $4\frac{1}{2}$; 3p.
In muddy weather, does any dirt cling to the feet of the partridge pheasant or any other birds, especially waders; has some cases of such, but supposes it is rare; case concerns distribution of plants with small seeds; cut off any dirt-clogged feet and send them to CD.

1. Watermark provides lower endpoint for date. Upper endpoint of 1858 assumed because this topic is discussed in Darwin, *Origin* (1859), pages 362-63ff. which CD had finished proofing before October 11, 1859 (see "Darwin's Journal," 15). Year of 1858 chosen over 1857 because of black border on stationery.

160 [1858] Nov. 8th. [pmk. NO 9/ 58]

To James EGAN; Down (type 2)
ALS; 8 X 5; 4p. and env., add. [M. James Egan/ Hotel Queen of England/ Pesth/ Austria/ via Belgium.]

Has seen Egan's article in Gardeners' Chronicle (edited by [John] Lindley, a friend of CD), and knows that Egan is a member of an agricultural society; do Hungarian horses frequently have a dark stripe down the spine, sometimes also a stripe (sometimes double) on the shoulders (as on the ass), and sometimes cross stripes on the legs; do such stripes occur frequently on the shoulders, on the front or on the hind legs, or on both; what color are horses with such stripes; are stripes plainer in the foal or in full-grown horse; in England, stripes are on duns, but CD unsure of the color of the parents of these duns.

161 [1858] Nov. 25th [pmk. NO25/ 58]

To J[ames] EGAN; Down (type 2)
ALS; 8 X 5; 3p. and env., add. [Mr J. Egan/ Hotel Queen Victoria/ Pesth/ Austria]

Thanks for prompt inquiries and reply concerning striped horses [see previous letter, above]; would appreciate any other information, esp. regarding the foal; [John] Lindley [editor of the Gardeners' Chronicle] would like to hear occasionally from Egan on Hungarian horticulture and on "the climate which produces the Tokay wine & fine Hungarian tobacco."

162 [1858] Decr. 12th

To [Walter ELLIOT]; Down (type 2)
ALS; 8 X 5; 4p., end. [Charles Darwin/ Abt Kabutar namah[1]/ & marks on Horses.]

Thanks for note of October 28, for poultry paper, and for treatise on pigeons; treatise shows some specimens sent to CD by Elliot to be nearly a century old; treatise shows difference between an Oriental and a European mind; belated thanks for the [Kaseen?][2] fowls; all CD's fowls are in hands of "a very skilful man," [William Bernhard] Tegetmeier, for description, and shall be deposited in British Museum; glad Elliot is joining Linnean Society [elected January 20, 1859]; is willing to sign Elliot's membership certificate; has Elliot seen stripes on backs, shoulders, and legs of horses and on legs of donkeys; asks for specific details about horses with stripes.

1. "Kabutar namah" is Hindustani for "pigeon reverence" or "pigeon salutation".
2. "Kaseen" may be CD's spelling of the Hindustani "kashin", which means "large" or "copious". See Duncan Forbes, A Dictionary, Hindustani and English..., new edition (London: Wm. H. Allen & Co., 1859).

163 [1859] March 28th [wmk. 1857]

To [Charles] LYELL; Down (type 2)
ALS; 4p. and enc. @ 8 X 5, 2p. @ 6¼ X 3 3/4; 6p. and 1p. enc.

Printed in full, with minor changes: Life and Letters, II, 151-52. Enclosure printed on page 152.

164 [1859] March 30th [end. March 30/ 1859; pmk. MR31/ 59]

To Ch[arles] LYELL; Down (type 2)
ALS; 8 X 5; 4p. and env., add. [Sir Ch. Lyell/ 53 Harley St/ London/ W.], end. [C. Darwin/ March 30/ Origin/ Of Species/ 1859]

Printed, with minor changes and minor omissions: Life and Letters, II, 152-53. At end of letter is: Emma [Wedgwood Darwin] goes to London for two or three days on Friday [April 1] and will visit Lyells on Saturday morning.

165 [1859] June 21st

To [Charles] LYELL; Down (type 2)
ALS; 8 X 5; 4p.

Printed in full, with minor changes: Life and Letters (seventh thousand revised, 1888), II, 159.

166 [1859 June][1] 28th

To [Charles] LYELL; Down (type 2)
ALS; 8 X 5; 2p.

Printed in full, with minor changes: More Letters, I, 125 (letter 77).

1. This month and year were provided without explanation or justification with the printed version. The month might be a bit early in the year.

167 [1859] Sept. 2d.

To [Charles] LYELL; Down (type 2)
ALS; 10 X 8; 4p. (enclosure wanting)

Printed, with minor changes: Life and Letters (seventh thousand revised,[1] 1888), II, 163-65. p. 163, line 3, missing name is "Emma [Wedgwood Darwin]". At end of letter is: whole family is ill; when proofs are finished (14 or 20 days), will leave for two months of hydropathy and rest; regards to wife; "I have read some of [Joseph Dalton] Hooker's Introduction to Australian Flora,[2] & he gives up species in grand style."; "I enclose P.S. of letter from [Alfred Russel] Wallace lately received."

1. The first portion of the letter is also omitted in the first edition.
2. *The Botany of the Antarctic Voyage of H.M. Discovery-Ships Erebus and Terror in the Years 1839-1843...*, Pt. III: *Flora Tasmaniae*, 2v. (London: Lovell Reeve, 1860), I, i-cxxviii. This "Introductory Essay" is dated "November, 4, 1859" on page cxxviii; it was reprinted separately in 1859, and portions of it appeared in some contemporary journals. In it, Hooker announced his acceptance of CD's theory of evolution by natural selection.

168 [1859][1] Sept. 11th

To Miss BUTLER; Down (type 2)
ALS; 7 3/4 X 5; 4p.

Hears that Butler is wandering about Scotland, so presumes she cannot go to Ilkley, but invites her anyway, since CD may not be able to take family; "It would be...terrible to go into that great place & not know a soul. But if you were there I should feel safe & home-like."; CD's book [*Origin* (1859)] is so nearly finished that he "shall be a free man at the end of this month"; thinks he will be in Ilkley for three or four weeks, followed by a week or so in Down and then a few weeks at Moor Park.

1. CD's only visit to Ilkley occurred in 1859; see "Darwin's Journal," 15.

169 [1859] Sept 20th

To [Charles] LYELL; Down (type 2)
ALS; 7 3/4 X 5; 8p.

Printed, with minor changes: *Life and Letters*, II, 166-67. At end of letter is: ill; leaves for Ilkley on October 3 [*sic*; October 2], by which time CD will have finished the "last revises [of the proofs of *Origin* (1859)], index & all"; fears it is too late for the whale correction, but has written to inquire about it;[1] "In Lecture to R[oyal] I[nstitution] [Richard] Owen showed that he believed in whale."

1. See Darwin, *Origin* (1859), 303-04; cf. *ibid.* (1860), 304. Apparently Lyell's claim about a whale fossil in the greensand was incorrect, so CD had it removed.

170 [1859] Sept. 25th

To [Charles] LYELL; Down (type 2)
ALS; 7 3/4 X 5; 6p.

Printed, with minor changes and minor omissions: Life and Letters, II, 168-69. p. 168, line 1, add: "The sheet with the whale-case has been printed off." p. 169, line 15, change "Chapter VIII" to "Ch. XIII". At end of letter is: leaves for Ilkley on September 29, arriving October 1; remembrances to Lyell's wife.

171 [1859] Sept. 30th [end. Sept. 1859; pmk. SE23/ 59]

To Charles LYELL; Down (type 2)
ALS; 7 3/4 X 5; 7p. and fragment of env., add. [Sir Charles Lyell/ Shielhill/ Kirriemuir/ Scotland], end. [C. Darwin/ finding out what the problems/ were to be solved more/ difficult than solution/ Sept. 1859]

Printed in full, with minor changes: Life and Letters, II, 169-71.

172 [1859] Oct. 11th

To [Charles] LYELL; Ilkley Wells House/ Otley, Yorkshire
ALS; 8¼ X 6½; 22p.

Printed, with minor changes: Life and Letters, II, 208-15. p. 213, line 11, add: on primrose and cowslip, has whole case written out in eight to ten pages; evidence bears out CD's conclusions; had experimented on this, but illness stopped the work; suggests that [Charles James Fox] Bunbury perform experiments, gives instructions for same; "I am assured if you sow lots of Polyanthus seed (but then these ought to have been secured from cross, & if starved plants the better) & sown in poor soil, you will get sometimes primroses & [other times] cowslips." p. 214, line 30, underline twice the "per-" in "perverted".

173 [1859] Oct. 20th [end. Oct. 20th 1859; pmk. OC20/ 59]

To Charles LYELL; Wells Terrace/ Ilkley Otley/ Yorkshire
ALS; 7 3/4 X 4 3/4; 8p. and env., add. [Sir Charles Lyell/ at C.J.F. Bunbury's Esqr/ Mildenhall/ Suffolk], end. [C. Darwin, Oct. 20th 1859/ "Creation." & archetypal creature/ Selection sufficient./ Droughts do not annihilate/ species.]

Printed in full, with minor changes: Life and Letters, II, 173-75.

174 [1859] Oct. 25th

To [Charles] LYELL; Wells Terrace/ Ilkley, Otley/ Yorkshire
ALS; 8¼ X 6½; 8p.

Printed, with minor changes: Life and Letters, II, 176-77. At beginning of letter is: gives "very hypothetical" doctrine of [Peter

Simon_/ Pallas as it applies to domestic dogs, viz. that present races of domestic dogs were produced by domestication of wolf in one country, of fox in another, etc., and by subsequent crossing; American dogs have descended from three or four "aboriginally distinct" species, and Europeans from several others; "We believe that all canine species have descended from one parent"; unclear whether all or only some differences in present breeds originated since domestication; importance of period of gestation has been exaggerated; races of man a great difficulty; does not believe Pallas's or [Jean Louis Rodolphe_7 Agassiz's claim that there are several species of man; [Sepoy_7 mutiny in India "stopped some important enquiries" about man; "I do not attribute much effect to climate &c."; some plants migrated through tropical lowlands during the glacial period; is lame; hopes H[enry_7 Holland will not review [Origin (1859)_7 in Q. Rev. because Holland "is so presumptuous & knows so little."

175 [1859 October_7 31st Monday

To [Charles_7 LYELL; Wells Terrace, Ilkley Otley Y[orkshire_7
ALS; $8\frac{1}{4}$ X $6\frac{1}{2}$; 4p.

Printed in full, with minor changes: More Letters, I, 127-28 (letter 80).

176 [1859_7 Nov. 23d

To [Charles_7 LYELL; Ilkley Wells./ Otley Yorkshire
ALS; $8\frac{1}{4}$ X $6\frac{1}{2}$; 4p.

Printed in full, with minor changes: Life and Letters (seventh thousand revised, 1888), II, 228-30.

177 [1859 ?November_7 24th

To [Thomas Campbell_7 EYTON; Ilkley Wells House/ Otley, Yorkshire
ALS; $7\frac{1}{4}$ X $4\frac{1}{2}$; 2p.

Thanks for note; CD's book [Origin (1859)_7 will "horrify & disgust" Eyton; "several high authorities" approved of CD's theories "far more... than I expected"; would like book to which Eyton refers; admires Eyton's zeal in going to Hythe to drill.

178 [1859 November_7 24th [wmk. 59?_7

To [Charles_7 LYELL; Ilkley Wells House/ Otley Yorkshire
ALS; $7\frac{1}{4}$ X $4\frac{1}{2}$; 4p. and 1p. enc.

Printed in full, with minor changes: Life and Letters (seventh thousand revised, 1888), II, 233-34. The enclosure reads: "Erasmus [Darwin] says about my Book [Origin (1859)] 'In fact the a priori reasoning is so entirely satisfactory to me, that if the facts wont fit in, why so much the worse for the facts is my feeling.'!" It is printed on page 233.

179 [1859 November] 25th

To [Charles] LYELL; Ilkley Wells House/ Otley, Yorkshire
ALS; $7\frac{1}{4}$ X $4\frac{1}{2}$; 4p.

Printed, with minor changes: Life and Letters (seventh thousand revised, 1888), II, 235-36. p. 235, line 3, add: has added a sentence on pheasants crossing;[1] has discussed this fully in [Darwin, Natural Selection]; pheasant species mentioned by CD undoubtedly blend by crossing, but CD is unsure whether the crosses are "quite fertile" inter se; cannot say more on this or on mistaken instincts because MS is at Down; would tell of blunder regarding instinct of wood ant, but the story is too long.

1. Compare Darwin, Origin (1859), 253, to ibid. (1860), 253. Or see Peckham, Variorum Origin, 433; the added sentence is numbered "69.1:b".

180 [1859 November] 29th

To [Charles] LYELL; no location
AL, S by init.; $3\frac{1}{2}$ X $8\frac{1}{4}$; 2p. (enclosure wanting)
Printed, with minor changes: More Letters, I, 136-37 (letter 91).
At end of letter is: "About rattle-snake I meant to have added, suppose the bead at end of tail of Trigonocephalus not to be moulted at each exuviation & to grow bigger with each new skin."

181 [1859] Dec. 2d.

To [Charles] LYELL; Ilkley Wells H[ouse] Otley Yorkshire
ALS; $8\frac{1}{4}$ X $6\frac{1}{2}$; 4p.

Printed: Life and Letters (seventh thousand revised, 1888), II, 236-37. p. 237, line 4, add: "I see he [John Crawford] does not even give me credit for knowing anything about the wild Columbidae allied to the Rock-pigeon!" p. 237, lines 9/10, change "several notes from ----" to "second note from [John] Phillips". p. 237, line 12, add: "Can he [Phillips] be staggered & have the fear of Oxford before his eyes." p. 237, line 12, change "X. says he" to "[Thomas Henry] Huxley says Phillips". At end of letter is: "I wish there was any chance of [Joseph] Prestwich being shaken; but I fear he is too much of a catastrophist."

182 [1859 December 3] Saturday

To /Charles_7 LYELL; no location
ALS; 8¼ X 6½; 2p.

First portion printed, with minor changes: More Letters, I, 129 (letter 81). line 3, add: "for he /Robert FitzRoy_7 wrote to me the other day on population of world not having increased, & in his Voyages there is the pebble theory.[1]" At end of this portion is: "What a mixture of conceit & folly, & the greatest newspaper in the world, inserts it!" Next portion printed, with minor changes: Life and Letters (seventh thousand revised, 1888), II, 239.

1. Narrative of the Surveying Voyages of His Majesty's Ships Adventure and Beagle..., 3v. Vol. II: Proceedings of the Second Expedition, 1831-1836 (London: Henry Colburn, 1839).

183 /1859_7 Dec. 5th

To /?Jean Louis Armand de QUATREFAGES de Bréau_7[1]; Down (type 2)[2]
ALS; 7¼ X 4½; 4p.

Thanks for letter of November 19; is "delighted that we agree" in part regarding "mutability of species"; will read /Quatrefages's_7 correspondence if published; C/harles_7 Lyell, /Joseph Dalton_7 Hooker, /William Benjamin_7 Carpenter, /Thomas Henry_7 Huxley, and others agree with CD's views; whole first edition /of Darwin, Origin (1859)_7, 1250 copies, was sold first day; publisher is now printing 3,000 more copies; /Anne-Louise Swanton_7 Belloc planned to translate Origin into French but found it too technical; knows /Quatrefages_7 is too busy to translate the book, but could he find a publisher, translator, or eminent naturalist to act as editor; will send copy of "2nd & corrected Edition" to any translator.

1. Correspondent is clearly French; Quatrefages is the most likely candidate, since he is known to have written an early congratulation to CD for the publication of the Origin. See Life and Letters (seventh thousand revised, 1888), II, 234.
2. CD was actually in Ilkley; see "Darwin's Journal," 15.

184 /1859 December 10_7 Saturday

To /Charles_7 LYELL; Down (type 2)
ALS; 8¼ X 6½; 6p.

Printed: Life and Letters (seventh thousand revised, 1888), II, 239-40. At beginning of letter is: thanks for suggestions /for Origin (1860)_7; could not include references on imperfection of geological record /chapter IX_7; estimate of 100,000 years for Mississippi deposition is taken from an extract of /Laurent-Guillaume_7 De Kerninck /sic; de Koninck_7 or /Étienne-Jules-Adolphe Desmier de Saint-Simon, Vicomte_7 d'Archaic; will now be more assertive about estimate, since Lyell's estimate agrees;[1] on richness of Purbeck beds, will add "*for thickness of beds*";[2] "On Friday I had interview with Sir H[enry] Holland, & found him going immense way with us (ie all Birds from one)--good, as showing how wind blows."; thinks the bigger of Lyell's interesting celts might have been used by those Eskimos who did not have iron to cut holes in ice and to kill glacial elephant and rhinoceros. p. 239, line 1, change "interviews" to "interview", and missing name is "[Richard] Owen". p. 239, line 2, add: "*but please repeat nothing*. Under garb of great civility, he was inclined to be most bitter & sneering against me. Yet". p. 239, line 4, add: "He was quite savage & crimson at my having put his name with defenders of immutability. When I said that was my impression & that of others, for several had remarked to me, that he would be dead against me: he then spoke of his own position in science & that of all the naturalists in London, 'with your [Thomas Henry] Huxleys', with a degree of arrogance I never saw approached." p. 240, line 9, add: "in most sneering tone". p. 240, line 23, add: "Lastly I thanked him him [sic] for Bear & Whale criticism, & said I had struck it out.-- 'Oh have you, well I was more struck with this than any other passage; you little know of the remarkable & essential relationship between bears & whales.'...by Jove I believe he thinks a sort of Bear was the grandpapa of Whales!....[3] We parted with high terms of consideration; which on reflexion I am almost sorry for. He is the most astounding creature I ever encountered."

1. For Lyell's estimate, see Lyell, *A Second Visit to the United States of North America* (London: John Murray, 1849), II, 248ff, esp. 250 and 250n. For Archaic's estimate, see Archaic, "Sur les formations dites pélagiques, et sur la profondeur à laquelle ont dû se déposer les couches de sédiment," *Bull. Soc. géol. Fr.*, *14* (1842-1843): 517-25. For CD's passage, cf. *Origin* (1859), 284, and *ibid.* (1860), 284; or see Peckham, *Variorum Origin*, 481-82, sentences 49 and 50.
2. Cf. Darwin, *Origin* (1859), 303, and *ibid.* (1860), 304; or see Peckham, *Variorum Origin*, 508, sentence 193.
3. Cf. Darwin, *Origin* (1859), 184, with *ibid.* (1860), 184; or see Peckham, *Variorum Origin*, 333, sentence 98.

185 [1859] Dec. 20th. [end. Ansd. 11/1./59][1]

To [William Henry SYKES]; Down (type 2)
ALS; 7 3/4 X 5; 3p. and end. [C. Darwin, the Distinguished/ Naturalist, Ansd. 11/1./59]

Recommends [Edward] Blyth for position as naturalist on the China Expedition.

1. Sykes apparently erred in his endorsement date; see letter to Lyell, December 29, 1859, below.

186 [1859 December] 22d

To ⌈Charles⌋ LYELL; Down (type 2)
ALS; 7 3/4 X 4 3/4; 4p.

Printed, with minor changes and minor omissions: Life and Letters, II, 245-46. p. 246, line 8, add: "What a marvellous geological Noah's ark that fossil tree in N. America was!"

187 ⌈1859 December⌋ 27th ⌈end. Dec 27.1859; pmk. DE 27/ 59⌋

To C⌈harles⌋ LYELL; Down (type 2)
AL (incomplete); 8¼ X 6½; 4p. and env., add. ⌈Sir C. Lyell/ 53. Harley St./ London/ (W.)⌋, end. ⌈C Darwin/ references to Clift &/ Lyell & Darwins journl. on/ connexion of extinct & existing/ types in S. America./ Dec 27.1859⌋

Printed, with minor changes and minor omissions: More Letters, I, 133-34.(letter 87). At end of letter is: "I doubt whether ⌈Joseph Dalton⌋ Hooker will succeed (anyhow I tried & failed) in keeping parts of Tropics hot, whilst other parts...⌈incomplete portion ends here⌋"

188 ⌈1859 December⌋ 29th ⌈wmk. 1859⌋

To ⌈Charles⌋ L⌈YELL⌋; Down.
AL, S by init.; 8 X 5; 4p.

Encloses note about ⌈Edward⌋ Blyth, as requested; Lyell should write such a note; also wrote to Col. ⌈William Henry⌋ Sykes;[1] suspects that ⌈Thomas Henry⌋ Huxley is author of article in the Times;[2] has not read notice in Daily News; has received "civil note" from ⌈Robert⌋ Chambers, containing news of an abstract in Chambers's Journal;[3] surprised (and said so to ⌈Richard⌋ Owen) at passage in Lyell's book; has not alluded to supposed British ⌈Trias?⌋ Mammal; believes not that mundane glacial period destroyed "all Tropical production," but that tropical and temperate forms mingled together on the plains during that period, as do ⌈Joseph Dalton⌋ Hooker's Himalayan forms; could give a Mexican example; "Anything on earth I can do in giving references &c &c will be a real pleasure...."; received letter from ⌈James Dwight⌋ Dana, who is "quite disable in his head" from overwork and is resting in Florence.[4]

1. See letter to Sykes, December 20, 1859, above.
2. Huxley's review of Darwin's Origin appeared in the London Times on December 26, 1859; see also Life and Letters, II, 252-55.
3. ⌈Chambers⌋, "Charles Darwin on the Origin of Species," Chambers's Journal, 12 (July-December, 1859): 388-91.
4. See Daniel C. Gilman, The Life of James Dwight Dana (New York: Harper & Brothers, 1899), 177.

189 ⌈?1859⌋[1] Dec. 31 ⌈wmk. 1859⌋

To ⌈William Hallowes⌋ MILLER; Down (type 2)
ALS; 8 X 5; 4p.

Very kind of Miller to do work completely; at present, will only publish the general result; measurements at upper thick end of comb are "most trustworthy", since bees "economise every particle of wax" and thus skimp on border cells; is glad that CD not as wrong as feared, even if accuracy was a result of "mere chance"; CD's original rough measurements were made in middle of comb, but recent discouraging measurements were made on border.

1. This letter appears to follow letters 73 to 75 in More Letters, I, 121-124.

190 ⌈1860 January⌋ 4th

To ⌈Charles⌋ L⌈YELL⌋; Down.
AL, S by init.; 8 X 5; 2p.

Printed, with minor changes and minor omissions: Life and Letters, II, 260-61. p. 260, line 7, add: argument in review ⌈of Darwin, Origin (1859)⌋ in Saturday Review is confined to geology, but reviewer gives CD "some perfectly just & severe raps on Knuckles."[1] p. 261, line 2, add: ⌈John Gwyn⌋ Jeffreys sent letter with "nonsense" about non-migration of sea-shells. At end of letter is: will send long letter from H⌈ewett⌋ C⌈ottrell⌋ Watson which CD has not read.

1. See "Darwin's Origin of Species," Sat. Rev., 8 (1859): 775-76.

191 ⌈1860⌋ Jan.10th ⌈pmk. JA11/ 60⌋

To C⌈harles⌋ LYELL; Down (type 2)
ALS; $8\frac{1}{4}$ X $6\frac{1}{2}$; 5p. and env., add. ⌈Sir C. Lyell/ 53. Harley St./ London/ (W.)⌋, end. ⌈C. Darwin/ Mortality of/ children./ Man originally an/ hermaphrodite./ Blind genus of/ insect with wide range/ Man & Spencer's/ Psychology.⌋

Printed, with minor changes and minor omissions: Life and Letters, II, 264-66. At beginning of letter is: is answering Lyell point by point; "Parthenogenesis (p. 96) is nothing & I know not why I inserted it in list."[1] p. 265, lines 6/7, remove parentheses around "only vaguely". p. 265, line 23, add: "I am very sorry that ⌈John⌋ Lindley did not write in Gardener's Chronicle."; and, after "Andrew Murray", add: "(the entomologist & dabbler in Botany)".

1. Cf. Darwin, Origin (1859), 96; and ibid. (1860), 96. Or see Peckham, Variorum Origin, 185, lines 128 and 128:b.

192 ⌈1860⌋ Jan 14th ⌈pmk. JA 15/ 60⌋

To C[harles] LYELL; Down (type 2)
ALS; $8\frac{1}{4}$ X $6\frac{1}{2}$; 6p. and env., add. [Sir C. Lyell/ 53. Harley St./ London/ (W.)], end. [C. Darwin/ Domestic vars of dog &c not/ as Huxley says far eno'/ to be sterile/ Cave insects common to N./ America & Europe./ Hooker on New Zealand/ not united in post-plio-/ cene times with Australia.]

Printed, with minor changes: Life and Letters, II, 266. At beginning of letter is: now that CD knows that [Joseph Dalton] Hooker is its author, has reread review in Gardeners' Chronicle;[1] letter from [John Gwyn] Jeffreys is not worth sending, it says nothing about migration but refers to two papers, one on Testacea;[2] spoke too strongly about Jeffreys's non-migration views; thinks Jeffreys "far too narrow & decided" in his opposition to [Edward] Forbes; is convinced from littoral shells at Galapagos that such shells have great power of migration; one of grandest points in Hooker's Essay[3] is demonstration that New Zealand has not been even nearly continuously united with Australia in recent times. p. 266, line 14, add: "I agree with [Thomas Henry] Huxley that it is a difficulty about no ascertained varieties known to have been raised by man, being sterile together: varieties of same kind, I believe, not rarely prefer pairing together; I have fact on this head. But I think Huxley had not considered my discussion (p. 268 New Edit. [i.e. Darwin, Origin (1860), 268])"; sterility of varieties of Verbascum and of tobaccos are wonderful cases; subject of sterility is "profoundly mysterious"; has been reading Isidore [Geoffroy Saint-Hilaire's] Life[4] of [his father, Étienne] Geoffroy St. Hilaire, plus the latter's Principes;[5] thinks latter man was "a rather doubtful maintainer of change of species"; former man has written to CD that he [i.e. Isidore] is "a firm maintainer of such views" and has sent a publication to show this; on Cave insects, difficulty applies chiefly in case of America and Europe, over which the same seeing genera range; if Lyell knows of any miocene or pliocene fossil insect in North America, such fact would make Lyell's hint very valuable. p. 266, line 17, "E." is "Emma [Wedgwood Darwin]". At end of letter is: can Lyell suggest a German translator; Madame [Anne Louise Swanton] Belloc finds [Darwin, Origin (1859)] too difficult to translate into French, but CD has just received letter from Frenchman eager to translate.

1. See Gdnrs' Chron., December 31, 1859. For a recent reprint of part of the review, see David L. Hull, Darwin and His Critics: The Reception of Darwin's Theory of Evolution by the Scientific Community (Cambridge, Mass.: Harvard Univ. Press, 1973), 81-86.
2. Jeffreys, "On the Marine Testacea of the Piedmontese Coast," Ann. Mag. nat. Hist., 17 (1856): 155-88.
3. "Introductory Essay," in The Botany of the Antarctic Voyage of H. M. Discovery-Ships Erebus and Terror in the Years 1839-1843..., Pt. III: Flora Tasmaniae, 2v. (London: Lovell Reeve, 1860), I, i-cxxviii.
4. Vie, travaux et doctrine scientifique d'E. Geoffroy Saint Hilaire (Paris: Strasbourg, 1847).
5. Principes de philosophie zoologique... (Paris: Acad. des Sciences, 1830).

193 [?1860] Jan. 21st [wmk. 1859]

To [Jean Louis Armand de QUATREFAGES de Bréau]; Down (type 2)
ALS; 8 X 5; 4p.

Sorry [Quatrefages] has been ill; thanks for "great Kindness"; man wishing to translate Darwin, [Origin (1859)] is M. [Pierre Theodore Alfred] Talandier, Professor of French at Royal Military College at Sandhurst, who writes well and is clever; if Talandier cannot find a publisher or changes his mind, CD will write to [Quatrefages]; heard that morning from N[orth] America that Darwin, [Origin (1859)] is "exciting considerable attention there amongst the naturalists", but [Jean Louis Rodolphe] Agassiz is "very savage at it"; has [Henri] Milne-Edwards read his copy [of Darwin, Origin (1859)] and what is his reaction;[1] are [Quatrefage's] lectures on anthropology[2] published?

1. The small portion of this letter dealing with Milne-Edwards has been printed, with minor changes, in two locations: Life and Letters (seventh thousand revised, 1888), II, 235; and More Letters, I, 136 (letter 90).
2. "Museum d'Histoire Naturelle. Anthropologie. Cours de M. de Quatrefages," Rev. scient., Paris, 5 (1867-1868): 366-69, 431-38, 450-55, 495-503, 510-18, 528-36, 544-50, 559-64, 579-84, 592-600, 621-31, 655-64, 685-96, 707-12, 720-28, 730-44, and 751-60.

194 [1860 January] 29th [end. 1860/ Jan 30; wmk. 1859]

To ?; Down (type 2)
ALS; 8 X 5; 2p., end. [1860/ C. Darwin Esqre/ Jan 30]

Thanks for assistance; encloses bank draft; wants "Agassiz Index Generum--it is a one volume Book & distinct from the Nomenclator";[1] [Jean Louis Armand de] Quatrefage[s de Bréau] is sending a book to CD via corr.; please send volumes to 57 Queen Anne Street, Cavendish Square.

1. Apparently a reference to Agassiz, Nomenclatoris Zoologici Index Universalis... (Soloduri: Jent et Gassman, 1848). This was a one-volume supplementary fascicle to Agassiz's Nomenclator Zoologicus....

195 [1860] Feb. 4th [end. Feb. 1860; wmk. 1859]

To [Philip Lutley SCLATER]; Down (type 2)
ALS; 8 X 5; 3p. and end. [5214/ C. Darwin/ Feb. 1860/ Abt Birds]
(enclosure wanting)

Thanks for list of Galapagos birds; if not too late, will include it in Darwin, Journal of Researches,[1] and will correct Darwin, Origin;[2] the wren is Sylvicola aureola, figured in [Darwin], Zoology of the Voyage of the Beagle;[3] have assumed it is confined to the archipelago; is surprised that Zenaida [galapagoensis] probably occurs on mainland; reiterates need for examples of variation of "abnormal parts" of birds, encloses list of examples of same.

1. See Darwin, Journal of Researches (1860), vii and 378-81.
2. Cf. Darwin, Origin (1860), 391, and ibid. (1861), 422. Or see Peckham, Variorum Origin, 620, lines 61 through 61:c.
3. Darwin, Zoology of the Voyage of the Beagle ([1838]-1843), Pt. III: Birds, by John Gould (1838-1841), 86 and Plate XXVIII.

196 [1860 February] 12th [wmk. 1859]

To [Charles] LYELL; Down (type 2)
ALS; 8 X 5; 4p.

Printed, with minor changes: Life and Letters, II, 280. At beginning of letter is: sends letters from [Heinrigh Georg] Bronn, from Asa Gray, and from [Charles James Fox] Bunbury;[1] has told Bunbury that undulatory theory of light is not a "vera causa"; on glacial distribution, Gray, in his letter, put CD's name before that of [Edward] Forbes, but Forbes deserves priority because he published first, even if CD had written out the notion earlier; will send first part of Gray's "excellent Review"[2] and notice by Bronn;[3] has just heard that, and is pleased that, Bronn will superintend the German translation of Darwin, Origin (1859); leave letters at home of Erasmus [Alvey Darwin]; will be in London "in fortnight". p. 280, line 2, add: "he hardly gave idea of my notions".

1. See Life and Letters, II, 268-73 and 276-80.
2. "Review of Darwin's Theory on the Origin of Species by Means of Natural Selection," Am. J. Sci. (Silliman's J.), 29 (1860): 153-84.
3. Neues Jb. Miner. Geol. Paläont., 1860; 112-16; for a recent English translation, see David L. Hull, Darwin and His Critics: The Reception of Darwin's Theory of Evolution by the Scientific Community (Cambridge, Mass.: Harvard Univ. Press, 1973), 118-25.

197 [1860] Feb. 14th [end. Feb. 1860; wmk. 1859]

To [Philip Lutley SCLATER]; Down (type 2)
ALS; 8 X 4 3/4; 2p. and end. [5208/ C. Darwin/ Feb. 1860/ About Birds]

Thanks for valuable information [see letter to Sclater dated February 4, 1860, above]; asked G[eorge] R[obert] Gray about Otus [galapagoensis] and Zenaida [galapagoensis];[1] do Strix punctatissima or Pyrocephalus nanus [on the Galapagos] differ in any degree whatever in size and duskiness from same species on mainland?[2]

1. Darwin, Zoology of the Voyage of the Beagle ([1838]-1843), Pt. III: Birds, by John Gould (1838-1841), 32-33, 115-16, and Plates III and XLVI.
2. Ibid., 34-35, 45-46, and Plates IV and VII.

198 [1860 February] 15th [and] 16th

To [Charles] LYELL; Down (type 2)
AL (incomplete); $8\frac{1}{4}$ X $6\frac{1}{2}$ (enclosures smaller); 4p. and 2p. enc. and 1p. enc.

Printed, with minor changes: <u>Life and Letters</u>, II, 284-86. At beginning of letter is: thanks for news; [?Auguste] Bravard's discoveries are magnificent, especially fact of "Palaeotherium Paranense" taken with "Nebraska Palaeotherium"; Bravard has sent two Spanish pamphlets with "strange geological doctrine, of whole enormous Pampean deposit being a <u>subaerial</u> deposit"; Bravard disputes unconvincingly the coembedment of Bahia Blanca fossils and recent shells; whole skeletons, including kneecap, cannot wash from one formation to another. p. 285, line 11, add: "The expression 'coincidence' in time & space between new & old species is unfortunate, as he believes, as we do, that new species are very <u>slowly</u> formed." p. 285, line 14, add: believes aberrant Anoa, or so-called antelope, is really small buffalo; work out interesting fact of Loess Man belonging to peculiar group of men; remember that fossil monkey was very manlike in middle Tertiary; will send Asa Gray's review when received from [Joseph Dalton] Hooker.[1] p. 285, line 23, incomplete letter in possession of APS ends at "I wish". First enc. reads as follows: "Many thanks for [Charles James Fox] Bunbury letter received this morning & for your note. I doubt whether I use term Natural Selection more as a Person, than writers use Attraction of Gravity as governing the movement of Planets &c but I suppose I could have avoided the ambiguity. 16th". Second enc. is: hopes Lyell discovers for what the <u>great</u> celts were used, since this bears on civilisation of old natives; [John Stevens] Henslow will visit celt beds in France during spring.

1. See letter to Lyell dated February 12, 1860, above.

199 [1860 February] 18th [and 19th; end. Feby 20./ 1860; pmk. FE20/60; wmk. 1859]

To C[harles] LYELL; Down (type 2)
ALS; 8 X 5; 10p. and fragment of env., add. [Sir C. Lyell/ 53. Harley St./ London/ (W.)], end. [C. Darwin/ Asa Gray's/ review of-/ Harveys monstrosity/ in Bigonia/ Bronn/ Feby 20./ 1860]

Printed in full, with minor changes: <u>More Letters</u>, I, 140-43 (letter 95). p. 142, line 19, change "polypus" to "polyps". p. 142, line 28, change "a future" to "any future". p. 143, line 13, change "at most" to "almost".
Small portion also printed in <u>Life and Letters</u>, II, 275.

200 [1860] Feb. 23d

To [Charles] LYELL; Down (type 2)
ALS; $8\frac{1}{4}$ X $6\frac{1}{2}$; 6p.

Printed, with minor changes: Life and Letters, II, 289-91. p. 289, line 10, add: [Joseph Dalton] Hooker will answer [William Henry] Harvey's notice [in Gardeners' Chronicle] if [John] Lindley [the editor] will permit;[1] Lyell can see this answer when at Down [March 9-13]; on issue of abrupt changes, case of Aspicarpa, like that of the differences between outer and inner florets of compositous and umbelliferous plants, is important case of modification of very important characters by correlation of growth, but it is not a case of abrupt origin of new forms; has tried to find cases of the latter but found only one "apparent case" in the Campanulaceae; concerning animals, besides case of monstrous goldfish,[2] has case of monstrous eels examined by [Jean Louis Rodolphe] Agassiz, but is unsure of this case; "On the whole I still feel excessively doubtful whether such abrupt changes have more than very rarely taken place." p. 290, line 12, add: does not understand [Heinrich Georg] Bronn's quote about Lyell; has read of infusorial experiments in Paris rejected as inaccurate by [Jean Louis Armand de] Quatrefage[s de Bréau]; similar old experiments were performed more carefully in Germany, with negative results.

1. See: Life and Letters, II, 274-76; and Gdnrs' Chron., February 18, 1860.
2. See More Letters, I, 141 (letter 95).

201 [1860] Feb. 25th [wmk. 1859]

To [Charles] LYELL; Down (type 2)
ALS; 8 X 5; 4p.

Printed, with minor changes: Life and Letters, II, 291. At beginning of letter is: glad to hear about [Richard] Owen; expects "many & bitter sneers" from Owen; glad Lyell used the same safe and true argument as [Herbert] Spencer's to the Bishop [?Samuel Wilberforce]. At end of letter is: sends portion of letter from [Heinrich Georg] Bronn which shows that Bronn is thinking more about [Darwin, Origin (1859)]; Bronn will translate it himself; has had letter from Sir W[illiam] Jardine, who opposes CD, but attack on CD's ornithological accuracy by Jardine is worthless; Jardine says Andrew Murray has read paper against CD;[1] does not know if degraded Aspicarpa flowers make fruit, but some other degraded flowers are abnormally fertile; agrees with criticisms of H. Spencer; has read Spencer's essay on population in which Spencer "publishes such dreadful hypothetical rubbish on the nature of reproduction."[2]

1. "On Mr. Darwin's Theory of the Origin of Species," Proc. R. Soc. Edinb., 4 (1857-1862): 274-91.
2. "A Theory of Population, Deduced from the General Law of Animal Fertility," Westminster Review, 57 (January-April, 1852): 468-501.

202 [1860] March 4th.

To [Joseph LEIDY]; Down (type 2)
Typed copy of ALS;[1] 11 X 8½; 2p.

Thanks for note of December 10 and valuable bundle of Leidy's publications; Leidy's palaeontology is highly regarded; Leidy's support is especially valuable because most palaeontologists "despise my work"; all older geologists except [Charles] Lyell are even more vehemently opposed; several younger geologists, however, support CD, especially on imperfection of geological record; is delighted that Leidy has evidence to support CD because, although CD himself will continue to work on the subject of evolution, "the sole way of getting my views partially accepted will be by sound workers showing that they partially accept them. I say partially, for I have never for a moment doubted, that though I cannot see my errors, that much in my Book [Origin (1859)] will be proved erroneous."

1. Original of this letter is at the Academy of Natural Sciences of Philadelphia, Philadelphia, Pennsylvania.

203 [1860 March] 12th [wmk. 1859][1]

To [Charles] LYELL; Down.
ALS; 8 X 5; 4p.

Printed in full, with minor changes: Life and Letters, II, 295.

1. This letter might be dated incorrectly by CD, since Lyell's species notebooks show him visiting CD until March 13, and this latter date seems like a more natural one for this letter. See Leonard G. Wilson, ed., Sir Charles Lyell's Scientific Journals on the Species Question (New Haven: Yale Univ. Press, 1970), 363 and 398. For further discussion of Greeks, see ibid., 364-65.

204 [1860 late March-early April][1]

To [Charles] L[YELL]; no location
AL, S by init.; 4½ X 7; 2p. (enclosure wanting)

Lyell is kind; hopes enc. will do; did not mention personal qualifications, of which CD is ignorant; could not allude to "precedent under [Robert John Eden,] L[or]d Aukland" because CD knows nothing about such precedent, not even whether there was a naturalist [?on the expedition]; will write to [Edward] Blyth in afternoon.
Next portion printed in full, with minor changes: Life and Letters, II, 297-98.

1. This letter must have been written after the publication of Sedgwick's review of Darwin, Origin (1859), which appeared in the Spectator on March 24, 1860, and it must have been written before the letter to Asa Gray dated April 3, [1860], printed in Life and Letters, II, 296-97,

because this letter claims that CD and Lyell agree that Sedgwick is the *Spectator*'s reviewer.

205 [?1860][1] April 7th [wmk. 1859]

To [Albert] WAY; Down (type 2)
ALS; 8 X 5; 4p.

Wants to know history of strongly marked domestic breeds of animals; thinks archaeologists may know when dray horses were first recorded; does Way know any archaeologists who can help on this point; should CD ask for help in Notes and Queries, "though that is a periodical I have no means of seeing"; "Eheu Eheu, the old Crux Major days are long past."[2]

1. The combination of type of Down address variant used on this letter with a watermark of 1859 occurred, apparently, only between late 1859 and mid-1860.
2. See Barlow, ed., *Autobiography*, 62-63. *Panagaeus crux-major*, a species of beetle, was collected by CD and Way while they were classmates at Cambridge.

206 [1860] Apr. 10th [wmk. 1859]

To [Charles] LYELL; Down (type 2)
ALS; 8 X 5; 8p.

Printed, with minor changes: *Life and Letters*, II, 300-01. p. 300, line 5, add: "I have got [the first number of the short-lived journal called] 'The Future' [published April, 1860], but cannot clearly make head or tail of it." p. 300, line 28, missing name is [Richard] Owen. p. 301, line 8, add: "makes me say that the dorsal vertebrae of pigeons vary & refers to page where the word dorsal does not appear. Sneers at my saying a certain organ is the branchiae of Balanidae; whilst in his own 'Invertebrata'[1] published before I published on cirripedes, he calls them organs without doubt branchiae." p. 301, line 10, missing name is Owen. At end of letter is: "How hard it is to please everyone. I told [Joseph Beete] Jukes that I sh[oul]d leave out in any next Edition [of Darwin, *Origin*] about the Weald,[2] & he demurred greatly & said 'he almost fancied he had written [it] himself,' as he bravely told [Roderick Impey] Murchison."

1. Richard Owen, *Lectures on the Comparative Anatomy and Physiology of the Invertebrate Animals...* (London: Longmans, Brown, Green, and Longmans, 1843), 158.
2. Cf. Darwin, *Origin* (1860), 285ff, and *ibid*. (1861), 308; or see Peckham, *Variorum Origin*, 483-85, lines 57-71.

207 [1860][1] Ap. 13 [wmk. 1859]

To ⌈Heinrich Georg BRONN⌋; Down (type 2)
ALS; 7 3/4 X 4 3/4; 1p.

Thanks for ⌈Bronn's⌋ "Untersuchungen uber ⌈sic⌋ die Entwickelung" and two copies of "Morphologische Studien," just received; will send extra copy of latter to "some good man"; thanks also for portrait.

1. Date and correspondent are obvious from the contents, as the following indicates: Bronn's books are Untersuchungen über die Entwickelungs-Gesetze der organischen Welt... (Stuttgart: E. Schweizerbart'sche Verlagshandlung und Druckerei, 1858) and Morphologische Studien über die Gestaltungs-Gesetze der Naturkörper überhaupt... (Leipzig and Heidelberg: C. F. Winter'sche Verlagshandlung, 1858). CD offered the extra copy of the latter book to Thomas Henry Huxley; see More Letters, II, 232 (letter 566).

208 ⌈1860⌋ Ap. 15th ⌈wmk. 1859⌋

To ⌈Charles⌋ LYELL; Down (type 2)
ALS; 8 X 5; 6p.

Printed in full, with minor changes: Life and Letters, II, 303-04.
p. 303, line 4, after "noticing", add: "⌈Richard⌋ Owen's".

209 ⌈1860⌋ Ap. 27th/ 28th ⌈wmk. 1859⌋

To ⌈Charles⌋ LYELL; Down (type 2)
ALS; 8 X 5; 8p.

Printed, with minor changes: More Letters, II, 30 (letter 403). At beginning of letter is: thanks for ⌈John Strong⌋ Newberry;[1] "the noncomittal ⌈sic⌋ men do not always most help a science"; sent "clever review" by ⌈Antoine Auguste⌋ Laugel[2] & address of President of Tyneside Naturalists;[3] regarding dogs, although the case is doubtful, CD favors multiple origins, but prefers not to commit himself until he can weigh all evidence; given the volume of discussion since ⌈Peter Simon⌋ Pallas, "I do not at all believe that ⌈Richard⌋ Owen did not know perfectly well some of the wild Canidae to which I alluded"; dogs of the world mingle the bloods of the European wolf, two distinct North American wolves, probably the Guyana dog or wolf, and probably (according to Pallas and Isidore Geoffroy St. Hilaire) several wild races of jackal; will compile facts in present volume,[4] but not in Origin; ⌈John William⌋ Dawson's remark on variability of Canidae will be hard to prove;[5] has received Lyell's budgets; thanks for drawings, sent on to Paris; supposes that ⌈William Benjamin⌋ Carpenter calls Vertebrata a class and birds an order, but this is unusual, and estimating the value of groups is "hopelessly difficult"; case of spitz dog is from ⌈Johann Matthäus⌋ Bechstein, ⌈Gemeinützige Naturgeschichte Deutschlands nach allen drei Reichen, 2nd ed., 4v. (Leipzig: Bey S. L. Crusius, 1801-1809), I, 638⌋; case of sheep and goat in Chile is "nearest case ⌈ever⌋ of reversion to one pure parent by repeated crosses"; such reversion is easy, and the number of generations needed to do it with various plants has been set by [Karl Friedrich von] Gärtner and perhaps [Joseph Gottlieb] Kölreuter; such reversion has been effected with Phasianus colchicus and P. versicolor; but Lyell refers to reversion only when hybrids are bred inter se, so Lyell's case seems "very wonderful & improbable"; is interested in Lyell's closing remarks.

1. Probably "Notes on the Ancient Vegetation of North America," Am. J. Sci. (Silliman's J.), 29 (1860): 208-18.
2. See Revue des Deux Mondes, April 1, 1860.
3. Probably in Transactions of the Tyneside Naturalists' Field Club, 4 (1858-1860).
4. Darwin, Variation under Domestication, 2v. (1868), esp. chap. 1.
5. Dawson, Archaia (Montreal: B. Dawson & Son, 1860). Lyell was reading this book at the time; see Life of Lyell, II, 332.

210 [1860] May 4th [wmk. 1859][1]

To [Charles] LYELL; Down (type 2)
ALS; 8 X 5; 6p.

Printed, with minor changes: Life and Letters, II, 261-62. At beginning of letter is: has been promised, and will send, an arrowhead found in peat by Col. Erskine in Aberdeenshire, which was found with many others in one place "where there were stones pitted for the manufacture"; similarly, John [William] Lubbock [Baron Avebury] says flints in France are found in such vast numbers in peat that M. [Jacques] Boucher [de Crèvecoeur] de Perthes said Lubbock might take as many as he liked; these facts remove great difficulty in case of gravel-bed celts, viz. their large numbers; hopes Lyell will return to France; it took sketch by Lubbock to make positions of celts clear to CD; case deserves months of work; will keep [John Strong] Newberry's interesting paper [see preceding letter, above]; pleased at how strongly Newberry asserted existence of American continent since Palaeozoic times; suspects CD's "crude notion" of cause for our ignorance of pre-Palaeozoic deposits will be shown to be true. p. 261, line 8, add: look at Spirifers arranged by [John William] Salter.[2]

1. Printed version is incorrectly dated January 4, 1860.
2. See Life and Letters, II, 366-67.

211 [1860 May] 8th [wmk. 1859]

To [Charles] LYELL; Down (type 2)
ALS; 8 X 5; 6p.

Printed, with minor changes: Life and Letters, II, 306-07. At beginning of letter is: Lyell's letter contained "much news"; did not know that, despite efforts of [Roderick Impey] Murchison, the Cambrian, or [Joachim] Barrande's primordial, has been separated from the Lower Silurian; sorry that CD shall not be in London [?to attend meeting of Geological Society of London] on 16th because CD wanted to hear papers;[1] has written to [John William] Lubbock, [Baron Avebury], about meeting; will stay at home because daughter Etty [Henrietta Emma Darwin Litchfield] has remittent fever [malaria]. p. 306, line 6, add: concerning aster, remembers paper by Asa Gray and another that gives cases of two forms specifically distinct but "perfectly united by intermediate varieties or links." p. 307, line 5, add: "I do not suppose that this is much of honour; but".

1. G. P. Wall, "On the Geology of a Part of Venezuela and of Trinidad," Q. Jl geol. Soc. Lond., 16 (1860): 460-70; and E. Lartet, "On the Coexistence of Man with Certain Extinct Quadrupeds...," ibid., 471-79.

212 [1860] May 18th [wmk. 1860]

To [Charles] LYELL; Down (type 2)
ALS; 7 3/4 X 5; 8p.

Printed: Life and Letters, II, 308-09. p. 308, line 7, add: Thanks for letter of 15th; new facts about man are interesting; [Thomas Taylor] Lewis takes account of rabbit and hare from Isidore Geoffroy Saint-Hilaire, but CD did not see original account; would like to ascertain origin of original hybrids because there is an old variety called "Hare-rabbit"; gives similar case of "pheasant-fowl" hybrids that were really varieties; wants to see [Hermann] Schaafhausen's pamphlet on natural selection;[1] has ordered "Canadian Mag."[2] p. 309, line 5, add: ill health and interruptions slow CD's progress; "I can very plainly see, as I lately told [Joseph Dalton] Hooker, that my Book [Origin (1859)] would have been & [would] be a mere flash in the pan, were it not for you, Hooker & a few others." At end of letter is: daughter Etty [Henrietta Emma Darwin Litchfield] is slightly better, has been ill three weeks; has read [?review of Darwin, Origin (1859)] by [Dominique Alexandre] Godron and found it commonplace, in contrast to "capital paper on Means of Distribution";[3] has received eight-page poem, in unknown hand, "quizzing & lauding" CD and his [Origin]; in case CD had not mentioned before, some time ago a Manchester newspaper published a "very good & long quiz...showing that I had proved that 'might was right' was the universal law of nature."[4]

1. "Über Beständigkeit und Umwandlung der Arten," Verh. naturh. Ver. preuss. Rheinl., 10 (1853): 420-51.
2. This refers, perhaps, to John William Dawson, "Review of 'Darwin on the Origin of Species by Means of Natural Selection,'" Can. Naturalist, 5 (1860): 100-20.
3. "Considérations sur les migrations des végétaux et spécialement de ceux qui, étrangers au sol de la France, y ont été introduits accidentellement," Mém. Acad. Stanislas, 1853: 329-67.
4. See Life and Letters, II, 262.

213 [1860] May 22d [wmk. 1860]

To [Charles] LYELL; Down (type 2)
ALS; 7 3/4 X 5; 4p. (enclosure wanting)

Printed, with minor changes: More Letters, I, 151 (letter 102). At beginning of letter is: encloses letter from A[sa] Gray "received this morning"; Appletons [American publishers of Darwin, Origin] are gentlemen [for paying a royalty to CD], but payment for edition of 2,500 copies was not large;[1] sends Isidore G[eoffroy Saint-]H[ilaire] on hare-rabbit, page 222;[2] glad to know author of reportedly "savage" [review in] North British Review, not yet read by CD;[3] [CD's son] William at Norgate says medical review referred to by Asa Gray is [William Benjamin] Carpenter's;[4] sorry to trouble CD with [Adam] Sedgwick in Cambridge paper.[5] At end of letter is: Etty [Henrietta Emma Darwin Litchfield, CD's daughter] is the same, has been ill over three weeks; "What a fact about the Coral Land Shells!!!"

1. See: Historical Records Survey, Division of Professional and Service Projects, Works Projects Administration, Calendar of the Letters of Charles Robert Darwin to Asa Gray (Boston: Historical Records Survey, 1939; reprinted Wilmington, Delaware: Scholarly Resources, Inc., 1973), 25; and "Darwin's Journal," 15.
2. Histoire Naturelle Générale des Règnes Organiques..., 3v. (Paris: Victor Masson et Fils, 1854-1862), III, 222. This is probably the correct reference, even though the date of publication of volume 3 is too late (i.e. 1862); perhaps CD had a proof page.
3. [John Duns], "On the Origin of Species.... By Charles Darwin,...," N. Br. Rev., 32 [American ed., 27] (May, 1860): 245-63. For identity of the anonymous author, see Life and Letters, II, 311.
4. "The Theory of Development in Nature," Br. for. med.-chir. Rev., 25 (1860): 367-404.
5. "Professor Sedgwick on Darwin's Theory," Cambridge Chronicle and University Journal, Isle of Ely Herald, and Huntingdonshire Gazette, May 19, 1860, pp. 3-4. See also: More Letters, I, 149n; and Darwin and Henslow, 203-07 (letters 111-114).

214 [1860 June] 1st Friday night [wmk. 1860]

To [Charles] LYELL; Down (type 2)
ALS; 7 3/4 X 5; 6p.

Printed, with minor changes: Life and Letters, II, 314-16. At beginning of letter is: has sent for H[enry] Holland to aid local doctor in treatment of Etty [Henrietta Emma Darwin Litchfield, CD's daughter], whose fever is nearly five weeks old; has sent A[ndrew] Murray's reviews, containing "weak" speculations;[1] has not misrepresented [Jörgen Christian] Schiödte.[2] p. 315, line 5, change "....I have" to "I have ordered the Future,[3] & have". At end of letter is: sends Asa Gray's letter, despite its insignificance; never attended to gestation of dogs because domestic gestation periods are so variable, but has now begun inquiries; greyhound stud observed for CD had yielded periods of 60 or 61 to 65 or 66 days; lowness of rodents does not decrease fertility of hybrids, since even algae are subject to same laws as higher animals.

1. Francis Darwin claims that there were two Murray reviews in the same place (*Life and Letters*, II, 261n). I could find only one, as follows: *Proc. R. Soc. Edinb.*, *4* (1857-1862): 274-91.
2. See Darwin, *Origin* (1859), 138.
3. See *Future: A Journal of Philosophical Research and Criticism*, *1* (1860).

215 [1860] June 6 [wmk. 1860]

To [Charles] LYELL; Down (type 2)
ALS; 7 3/4 X 5; 8p.

Printed, with minor changes: *Life and Letters*, II, 317-19. At beginning of letter is: Etty [Henrietta Emma Darwin Litchfield, CD's daughter] is still ill; H[enry] Holland saw her on Sunday [June 3] and predicted long recovery; [Samuel] Haughton's review in "Dublin Mag. of Nat. Hist."[1] is "more coarsely contemptuous" than that of [John] Dunns [sic; Duns] in *N. Br. Rev.* [see CD to Lyell, May 22, 1860, above]; Haughton's "ignoring in his remarks on Bees' cells the almost exactly intermediate comb of Melipona" is unfair. p. 317, line 1, missing name is "[Haughton]". p. 317, line 18, add: "or Quinarianism." p. 318, line 18, add: "& likewise (*if you can spare*) [Edward William] Binney on Coal[2] & [Herman] Schaafhausen or some such name on Natural Selection [see CD to Lyell, May 18, 1860, above]." p. 318, bottom line, add: in Darwin, *Origin* [(1859), 137], attributed blindness of cave animals *exclusively* to disuse, not selections of chance varieties, but was hasty about insects, overlooking bearing of fact that blind genus *Adelops* lives under moss *out* of caves; possibly also genus *Anophthalmus* (a "Carabidous" beetle) was blind and "extra-cavernal"; "It seems not unlikely that a blind insect would be less inconvenienced in dark cave than other insects, & would become tenant"; several passages in review by [Andrew] Murray [see previous letter, above] were unclear. p. 319, line 4, add: "[Dominique Alexandre] Godron puts well [see CD to Lyell, May 18, 1860, above] the little effect of *climate*, which...becomes stronger...on my mind. I do not say confidently food."

1. "On the Form of the Cells Made by Various Wasps, and by the Honey Bee; With an Appendix on the Origin of Species," *Proc. nat. Hist. Soc. Dubl.*, *3* (1859-1862): 128-40.
2. "Sketch of the Drift Deposits of Manchester and Its Neighbourhood," *Mem. Manchr. lit. phil. Soc.*, *8* (1848): 195-234. See also: Leonard G. Wilson, ed., *Sir Charles Lyell's Scientific Journals on the Species Question* (New Haven: Yale Univ. Press, 1970), 404 and 480.

216 [1860 June] 14th [wmk. 1860]

To /̅Charles_7 LYELL; Down (type 2)
ALS; 7 3/4 X 5; 5p. and 1p. enc. @ $4\frac{1}{2}$ X $7\frac{1}{4}$ (other enclosures wanting)

Printed, with minor changes: Life and Letters, II, 319-20. At beginning of letter is: Etty /̅ Henrietta Emma Darwin Litchfield_7 is in slightly worse health; CD cannot walk; sends extract of letter from /̅Edward_7 Blyth, who is grateful to Lyell for help concerning Chinese expedition;[1] encloses letter from /̅William_7 Hopkins; fancies Hopkins is "horrified about man; I have told him that I thought man must be included under same category with animals"; returns four pamphlets; does not see much in Binney /̅ see previous letter, above_7. p. 319, line 5, add: "It is no wonder /̅Jean Louis Rodolphe_7 Agassiz denies varieties in animals, when he calls even the same forms in two distinct countries, two Species." At end of letter is: after much puzzling, does not know what A/̅ndrew_7 Murray meant /̅see CD to Lyell, June 1, 1860, above_7 by "Agassiz & ab ovo". End. reads: Blyth says there is account of flint tools found in ice in /̅Elizabeth Juliana Leeves_7 Sabine's translation of /̅Ferdinand Petrovich, Baron von_7 Wrangell's /̅i.e. Vrangel′'s_7 Voyage[2] on page 117 of introduction; this find relates to issues like mastodon found in ice, and is worth checking; Blyth thinks Eskimos, when first discovered, had no iron tools, and he refers to essay[3] by J/̅ohn_7 Richardson.

1. See the following letters, all abstracted above: CD to W. H. Sykes, December 20, 1859; CD to Lyell, December 29, 1859; and CD to Lyell, late March-early April, 1860. See also Loren C. Eiseley, "Charles Darwin, Edward Blyth, and the Theory of Natural Selection," Proc. Am. phil. Soc., 103 (1959): 154-55.
2. Narrative of an Expedition to the Polar Sea..., tr. ... by Mrs. Sabine, ed. by E. Sabine (London: J. Madden and Co., 1840).
3. "Esquimaux: Their Geographical Distribution," Edinb. new phil. J., 52 (1852), 322-23.

217 /̅1860_7 June 17th /̅wmk. 1860_7

To /̅Charles_7 LYELL; Down (type 2)
ALS; 7 3/4 X 5; 4p.

Printed, with minor changes and minor omissions: More Letters, I, 154 (letter 105). line 24, add: Etty /̅Henrietta Emma Darwin Litchfield_7 is slightly better in health, has been ill for exactly seven weeks.

218 /̅1860_7[1] June 18 /̅?wmk. 1860_7

To /̅Thomas Campbell_7 EYTON; Down (type 2)
ALS; 7 3/4 X 5; 1p.

Please send to CD any authentic cases of duration of gestation in dogs; hounds, otter-hounds, or any breed will be acceptable.

1. Portion of a watermark on this letter appears to be 1860. Also, dimensions of stationery and Down address used fit a dating of 1860. Finally, CD was studying gestation of dogs for the first time in June, 1860; see CD to Charlex Lyell, June 1, 1860, above.

219 [1860 June] 20th [wmk. 1860]

To [Charles] LYELL; Down (type 2)
ALS; 7 3/4 X 5; 4p.

Printed in full, with minor changes: More Letters, I, 155-56 (letter 106). p. 155, line 8, change "surely we" to "surely as we".

220 [1860 June] 25th [wmk. 1860]

To [Charles] LYELL; Down (type 2)
ALS; 7 3/4 X 5; 4p. (enclosures wanting)

Printed, with minor changes: More Letters, I, 155n. At beginning of letter is: encloses authentic arrowheads, with note describing location of discovery, which were sent to CD by Mrs. Moir, mother-in-law of Col. Erskine.[1] line 4, questionable word is definitely "record". At end of letter is: because of bad stomach, will give up, reluctantly, his plans to attend [B.A.A.S. meeting at] Oxford [June 27-July 4]; will leave Thursday [June 28] for one week of water cure at Dr. [Richard James] Lane's, Sudbrook Park, Richmond, Surrey; will not stay longer because Etty [Henrietta Emma Darwin Litchfield, CD's daughter], while better, is still ill; thanks for letter just received; returns unimpressive letter from [John William] Dawson;[2] wishes to borrow Dawson's review [of Darwin, Origin (1859)]; on Dawson's letter, it "would be insanity to compare evidence of organic change with geological change, at present, as far as strength of evidence goes. But what inches of elevation on coast of Sweden are to great mountains so are the numerous varieties & endless doubts what to call species & what varieties, to undoubted species. I entirely deny that there is no evidence of change. But time alone will bring naturalists round, when they find that they can explain many facts on such views as mine, & cannot on view of creation."

1. See CD to Lyell, May 4, 1860, above.
2. For Lyell's comments on Dawson's letter, see Leonard G. Wilson, ed., Sir Charles Lyell's Scientific Journals on the Species Question (New Haven: Yale Univ. Press, 1970), 457-58.

221 [1860 July] 5th. Thursday [wmk. 1860]

To [Charles] LYELL; Sudbrook Park/ Richmond
ALS; 7 3/4 X 5; 4p., plus 1p. fragment of AD by Lyell @ 6 X 5

Thanks for note; glad Lyell is going to Amiens; while there, please explain high and low dispersion of flint gravel; also glad Lyell to investigate post-glacial period; Etty [Henrietta Emma Darwin Litchfield, CD's daughter] moved to Hartfield on Tuesday; leaves on Saturday; glad CD did not attempt [B.A.A.S. meeting at] Oxford; "[Thomas Henry] Huxley, [Joseph Dalton] Hooker & J[ohn] Lubbock (as I am pleased to hear) seem to have stuck up for modification of Species like Trojans"; Asa Gray also goes on fighting for CD's theory;[1] thinks "we shall in long run conquer"; likes [article by William] Hopkins in Fraser's Magazine,[2] but regrets "soul-discussion"; difficulties alone "make a very damaging review"; Lyell's facts convinced CD for first time that hare-rabbits are hybrids; could not confirm this before.[3] Enc. is entitled "C. Darwin/ On Species & Creation" and is merely a list of such headings as "Bermuda & Madeira Birds why like Continental--" and "Primrose & cowslip".

1. ?See Historical Records Survey, Division of Professional and Service Projects, Works Projects Administration, Calendar of the Letters of Charles Robert Darwin to Asa Gray (Boston: Historical Records Survey, 1939; reprinted Wilmington, Delaware: Scholarly Resources, Inc., 1973), 27-28.
2. "Physical Theories of the Phenomena of Life," 61 (1860): 739-52; and 62 (1860): 74-90.
3. See: CD to Lyell, May 18, 1860, above; CD to Lyell, May 22, 1860, above; and Darwin, Variation under Domestication (1868), I, 105 and 105n.

222 [1860] July 30th

To [Charles] LYELL; Hartfield
ALS; $10\frac{1}{2}$ X $8\frac{1}{4}$; 2p.

Printed, with minor changes: Life and Letters, II, 327-28. At beginning of letter is: have been at Hartfield three weeks because Etty [Henrietta Emma Darwin Litchfield, CD's daughter] is still ill; return home in few days [August 2]; H[enry] Holland saw Etty few days ago, predicted long recovery; has seen no one except [Joseph Dalton] Hooker for hour or two at Kew. p. 327, line 9, add: "considering [Richard] Owen's aid [by the way it seems generally admitted that Huxley smashed Owen at Oxford][1] [CD brackets]; it quizzes me really in capital style". p. 328, line 3, missing name is "Owen". p. 328, line 5, add: "Owen is really wonderfully clever in his malevolence."

1. Reference is to Thomas Henry Huxley at the B.A.A.S. meeting at Oxford from June 27 to July 4, 1860.

223 [1860] August 11th

To [Charles] LYELL; Down (type 2)
ALS; $10\frac{1}{2}$ X $8\frac{1}{4}$; 4p.

Printed, with minor changes: *Life and Letters*, II, 331-32. At beginning of letter is: thanks for letter; have been home about a week; Etty [Henrietta Emma Darwin Litchfield, CD's daughter] improving, but anxiety over her has interrupted CD's work. p. 331, line 11, missing name is [Richard] Owen. p. 331, line 14, add: A[sa] Gray "argued capitally" for CD at second discussion [of natural selection] before American Academy [of Arts and Sciences]; Owen sent copy of "one of his Reports, so he does not wish to come to quarrel with me." p. 331, line 20, add: Rudolf Wagner has published in Germany an abstract of [Jean Louis Rodolphe] Agassiz, *[An] Essay on Classification* [London: Longman & Co., 1859] in relation to '*Darwin ansichten*' and concludes that truth lies between CD and Agassiz, which "will make Agassiz savage"; [Thomas Henry] Huxley says [Karl Ernst Ritter] Von [*sic*; von] Baer [Edler von Huthorn] "goes a long way with us,...has spoken publicly &...will perhaps publish on subject." p. 331, line 24, add: there is a very good, geological, favorable third article [on natural selection] in *London [Quarterly and Holborn] Review*, author unknown. p. 332, line 7, add: Mrs. [Frances Harriet Henslow] H[ooker] and baby [Brian Harvey Hodgson Hooker] are at Worthing; latter is ill.

224 [1860] Aug 28th

To [Charles] LYELL; Down (type 2)
ALS; $10\frac{1}{2}$ X $8\frac{1}{4}$; 2p.

While in London lately, heard that Sir G[eorge] Grey found his wife [Harriet Spencer Grey] in bed with "Capt. Keppell"; saw [Hugh] Falconer, who spoke of "tiny new species of Elephant from Malta"; Asa Gray's review [i.e. part two of "Darwin on the Origin of Species"] in August *Atlantic Monthly* [6 (1860): 229-39] is excellent, argumentative;[1] Gray is "a first rate arguer" who "most completely understands the subject"; CD has been abused in Catholic journal; glad that "Rajah Sir J[ames] Brook" is well again; work going well, today finished dogs;[2] still believes dogs descended from "several wild stocks"; sent *Athenaeum* and *Quarterly Review*.

1. For a recent reprint and a full publication history of this review, see Asa Gray, *Darwiniana...*, ed. A. Hunter Dupree (Cambridge, Mass.: Belknap Press, 1963), 72n and 85-105.
2. Darwin, *Variation under Domestication* (1868), I, 15-43.

225 [1860] Sept. 1.

To [Charles] LYELL; Down (type 2)
ALS; $10\frac{1}{2}$ X $8\frac{1}{4}$; 4p.

Printed, with minor changes and minor omissions: *Life and Letters*, II, 334-37. p. 336, line 22, after "applies", add "perhaps". p. 336, line 29, change "clear" to "close".

226 [1860] Sept. 12th

To [Charles] LYELL; Down (type 2)
ALS; 10½ X 8¼; 4p.

Printed, with minor changes: *Life and Letters*, II, 339-41. p. 339, line 5, add: "Even if his mind had not been full of [thoughts about his trip to] Syria he could never have conjectured your precise line of thought." p. 340, line 13, add: cannot estimate number of species "'extinguished in a given time'"; passages on pages 168 and 313 [of Darwin, *Origin* (1859) or (1860)] are not contradictory, because "mere variability & variability taken advantage of & selected are widely different considerations"; gives example of rudimentary organs; has not been guarded enough in claim that Ammonites became extinct *relatively* suddenly compared to other families [see *Origin*, 321-22]; has alluded to much extinction and modification in great intervals between formations; thinks it striking that in southern Chile near Concepcion, there are apparently Tertiary beds with Ammonites and Baculites. p. 341, line 7, add: Darwins go to sea [Eastbourne] in about a week [Sept. 22].[1]

1. For Lyell's letters to CD during this period, see Leonard G. Wilson, ed., *Sir Charles Lyell's Scientific Journals on the Species Question* (New Haven: Yale Univ. Press, 1970), 467-69, 472-77.

227 [1860 September] 23d Sunday

To [Charles] LYELL; 15 Marine Parade/ Eastbourne
ALS; 8¼ X 6½; 10p. and 2 sketches

Printed, with minor changes: *Life and Letters*, II, 341-44. p. 341, line 6, add: has sent review by A[sa] Gray;[1] as *Ann. Mag. nat. Hist.* has printed [Jean Louis Rodolphe] Agassiz's article,[2] will ask its editor [William Jardine] in fairness to reprint Gray at CD's expense and with Gray's name attached; Gray's review good because it gives so much of [Francois Jules] Pictet [de la Rive]; "The Annals, I fear, have very small circulation"; misunderstood Lyell on types; mentions health of Etty [Henrietta Emma Darwin Litchfield, CD's daughter] and death of the [Thomas Henry] Huxleys's son [Noel Huxley]. p. 342, lines 12 to 14, add in left margin: "[Richard] Owen if he chose to attend to such view could work this out." p. 342, line 20, add: was silent because unsure that there was fossil rodent in Australia, but thought not; as to Australia's especial suitability for marsupials, "I have always thought it a gigantic hallucination of Owen.--not to mention Rodents"; dingo was wild long before South Australian volcanic outburst, and there are many marsupial species in Brazil; also, New Guinea, although humid, is tenanted by marsupials as exclusively as Australia; despite antiquity of dingo (referred to in dog MS[3]), thinks dingo introduced by man, and if so, this bears on antiquity of man; if dingo existed outside Australia then it is not aboriginal in Australia; [René Primevère] Lesson says same about dog of New Ireland, but Lesson not to be trusted; likes case of tree stump living by natural grafting of roots, wants reference on it; thinks case confined to Coniferae.[4] p. 342, line 27, add: [Jean Louis Rodolphe] Agassiz's remark in [Josiah Clark] Nott and [George Robins] Glidden[5] on coincidence of color alone being a fleeting character "does not go for much in his comparison of man & anthropoid apes."

1. See CD to Lyell, August 28, 1860, above.
2. "Prof. Agassiz on the Origin of Species," *Ann. Mag. nat. Hist.*, *6* (1860): 219-32.
3. Darwin, *Variation under Domestication* (1868), I, 15-43, in manuscript.
4. See also Leonard G. Wilson, ed., *Sir Charles Lyell's Scientific Journals on the Species Question* (New Haven: Yale Univ. Press, 1970), 476.
5. Louis Agassiz, "Sketch of the Natural Provinces of the Animal World and Their Relation to the Different Types of Man," in J. C. Nott and G. R. Glidden, *Types of Mankind; Or Ethnological Researches...* (Philadelphia: Lippincott, Grambo & Co., 1854), lxxv.

228 [1860 September] 26th [wmk. 1860]

To [Charles] LYELL; 15 Marine Parade/ Eastbourne
ALS; 7 3/4 X 5; 8p.

Printed in full, with minor changes: *More Letters*, I, 167-69 (letter 112).

229 [1860 September] 28 Friday Evening [wmk. 1860]

To [Charles] LYELL; 15. Marine Parade/ Eastbourne
ALS; 7 3/4 X 5; 8p. (slightly mutilated)

Printed, with minor changes: *Life and Letters*, II, 345-46. At beginning of letter is: mentions Lyell's letter; extinction of Ammonites is "a most singular fact" despite what Lyell says about great breaks in upper chalk; Lyell may obtain *Atlantic Monthly* at Trubners. p. 345, line 6, add: "but in very difficult points, &". p. 346, line 2, add: has asked A[sa] Gray where [Karl Ernst Ritter] Von [*sic*; von] Baer [Edler von Huthorn] makes statement about guinea pig, which is worthless unless there is new evidence about wild parent; denies that aperea of La Plata and southern Brazil is the wild parent stock; guinea pig was domesticated when America was discovered; von Baer has read [Darwin, *Origin* (1859)] approvingly; would keep [?i.e. breed] hybrid hare-rabbit himself, but still would not have evidence of hybridity of any specimens obtained from France; [Abraham Dee] Bartlett is correct to try to cross wild hare and rabbit, but he should try several races of rabbit.

230 [1860] Oct. 3d [wmk. 1860]

To [Charles] LYELL; 15 Marine Parade/ Eastbourne
ALS; 7 3/4 X 5; 11p.

Printed, with minor changes: More Letters, I, 169-72 (letter 113). p. 169, line 1, add: "The Dog MS is safe at Down." p. 171, line 37, add: Lyell's remarks on Ammonites, cuttlefish, and Hippurites are interesting;[1] will think about keeping the rabbits; Isidore G[eoffroy] S[ain]t Hilaire, of whom CD has very good opinion, only knows the case second hand; [Richard] Owen "sneers at [Geoffroy Saint Hilaire]; & I daresay he [?] is not [to] be trusted on Homologies";[2] do not trust Sclagenweit [i.e. Hermann Rudolf Alfred von Schlagintweit-Sakünlünski and Robert von Schlagintweit] about yaks;[3] "there are many reputed species (laying on one side question of fertility) not so distinct as negro & white man."

1. See Leonard G. Wilson, ed., Sir Charles Lyell's Scientific Journals on the Species Question (New Haven: Yale Univ. Press, 1970), 497.
2. For Geoffroy Saint Hilaire on homologies, see Stauffer, ed., CD's Nat. Selection, 298 and 298n.
3. Hermann and Robert Schlagintweit, "Notes on Some of the Animals of Tibet and India," Rep. Br. Ass. Advmnt Sci., 27 (1857), pt. 2: 106-08. See also Stauffer, ed., CD's Nat. Selection, 438 and 438n.

231 [1860 October] 5th Friday [wmk. 1860]

To [Charles] LYELL; 15 Marine Parade/ Eastbourne
ALS; 7 3/4 X 5; 6p.

Printed, with minor changes: More Letters, I, 172 (letter 114). line 8, change "creative" to "creation". At end of letter is: bats are "washed out of my head"; species in Madeira, collected by Mr. Mason [?], are European; does not remember about Palma specimens, given to [Robert Fisher] Tomes of Welford; did not get Azores species; "I had heard nothing of the sales of 'Origin' [i.e. Darwin, Origin] for months, & am much pleased to hear that the sale continues; this surprises me"; has not received second Atlantic article,[1] meant for [Francois Jules] Pictet [de la Rive]; has not received theological dialogue;[2] [Hugh] Falconer, who is only man who has facts correct, is critical of Isidore Geoffroy [Saint Hilaire]; will check on St. Helena concerning [? Geoffroy St. Hilaire's] conjecture about number of plants exterminated; St. Helena was woody in late periods; see Darwin, Journal of Researches; is "wasting time shamefully" on Drosera experiments which are "perverse & crooked"; Etty [i.e. Henrietta Emma Darwin Litchfield, CD's daughter] is gaining strength; [Heinrich Georg] Bronn has appended chapter of objections at end of translation;[3] Miss Ludwig has translated it for CD.

1. Asa Gray, "Darwin on the Origin of Species [Part II]," Atlantic Monthly, 6 (1860): 229-39. See CD to Lyell, Aug. 28, 1860, above.
2. "Discussion between Two Readers of Darwin's Treatise on the Origin

of Species, upon Its Natural Theology," Am. J. Sci. (Silliman's J.), 30 (1860): 226-39.
3. Darwin, Über die Entstehung der Arten... [German Origin], tr. into German by H. G. Bronn (Stuttgart: Schweizerbart'sche Verlag, 1860).

232 [1860] Oct 8. [wmk. 1860]

To [Charles] LYELL; 15 Marine Parade/ Eastbourne
ALS; 7 3/4 X 5; 6p. and 1p. enc. [broadside advertisement for C. R. Bree, Species Not Transmutable,... (London: Groombridge and Sons, 1860)], @ 7½ X 5

Printed, with minor changes: Life and Letters, II, 346-47. p. 347, line 19, add: has written to Down about missing reviews [see preceding letter, above]; believes southeast and southwest corners of Australia were islands, with latter older and more typical; [Joseph Dalton] Hooker has speculated on this in Introduction;[1] so have [Joseph Beete] Jukes and CD himself in review[2] of [George Robert] Waterhouse's [A Natural History of the] Mammalia [2v. (London: H. Ballière, 1846-1848)]; in Saturday's Athenaeum, Jukes answered capitally Sir [Henry] James's "wild speculations" on change of earth's axis.[3] p. 347, line 24, add: Miss L[udwig] says [Heinrich Georg] Bronn is very difficult German [see preceding letter, above]; has not heard of Bovey Coal Plants;[4] hopes [Charles James Fox] Bunbury will undertake them and that Bunbury's new position will not interfere with his science; does not know [Charles Robert] Bree, perhaps the son of Rev[eren]d [William Thomas] Bree, "a good miscellaneous observer of habits of all creatures...& Botanist."

1. See CD to Lyell, Sept. 2, 1859, above, esp. my note 2.
2. Ann. Mag. nat. Hist., 19 (1847): 53-56, esp. top of 56.
3. See More Letters, II, 140n.
4. See Life of Lyell, II, 346-47 and 349-50.

233 [1860] Nov. 20

To [Charles] LYELL; Down (type 2)
ALS; 8¼ X 6½; 4p. and 4p. enc.

First portion printed, with minor changes and minor omissions: More Letters, I, 461 (letter 351). At end of this portion is: Lyell's chapters must be difficult but are worth much labor; fears that Lyell's volume on geological history of man [i.e. The Geological Evidences of the Antiquity of Man (London: John Murray, 1863)] "will slip through your fingers"; does not think [Joseph Dalton] Hooker has criticized [Edward] Forbes; H[ewett] C[ottrell] Watson has abused Forbes in Cybele [Britannica; Or, British Plants and Their Geographical Relations, 4v. (London: Longman & Co., 1847-1852), I, 465-72], but it is not well done.
Next portion and enclosure printed, with minor changes and minor omissions:

Life and Letters, II, 349-51. p. 350, after signature, add: Etty [i.e. Henrietta Emma Darwin Litchfield, CD's daughter] goes on well, but is weak. p. 350, right column of enclosure, line 3, change "alludes to" to "attacks".

234 [1860] Nov. 24th [wmk. 1859]

To [Charles] LYELL; Down (type 2)
ALS; 4p. @ 8 X 5, 2p. @ 8 3/4 X 5½; 6p.

First portion printed, with minor changes and minor omissions: Life and Letters, II, 352, lines 1 to 10.
Next portion printed, with minor changes and minor omissions: Life and Letters, III, 319-20. At end of this portion is: Etty [i.e. Henrietta Emma Darwin Litchfield, CD's daughter] goes on well, but doctors say rapid progress is impossible; Drosera and dreadful illness for last six months has made progress on CD's larger book "almost nothing".
Next portion printed, with minor changes: Life and Letters, II, 352-53, beginning where first portion ended. p. 352, line 4 of this portion, missing name is "[Richard] Owen".

235 [1860] Nov. 25th

To [Charles] LYELL; Down (type 2)
ALS; 8¼ X 6½; 6p.

Printed in full, with minor changes: More Letters, II, 137-40 (letter 491). p. 139, line 29, change "the great" to "that great [an]".

236 [1860] Dec. 4th

To [Charles] LYELL; Down (type 2)
ALS; 8 X 6½; 5p.

First portion printed, with minor changes and minor omissions: More Letters, II, 140-41 (letter 492). **p. 141, line 10, add: "How far** to lump & split species is indeed a hopeless problem. It must in the end, I think, be determined by mere convenience." At end of this portion is: glad to hear that Lyell continues to "stir them up" at Zoolog[ical] Soc[iety of London].
Next portion printed, with minor changes and minor omissions: Life and Letters, II, 352n. At end of this portion is: there is an article[1] on Darwin, Origin, in Macmillan's Magazine; has not yet read [John] Phillips, [Life on the Earth, Its Origin and Succession (Cambridge: Macmillan and Co., 1860)].

1. Henry Fawcett, "A Popular Exposition of Mr. Darwin on the Origin of

Species," *3* (December, 1860): 81-92. For a recent reprinting, see David L. Hull, *Darwin and His Critics: The Reception of Darwin's Theory of Evolution by the Scientific Community* (Cambridge, Mass.: Harvard Univ. Press, 1973), 277-90. See also *Life and Letters*, II, 299.

237 [?1860-1861][1] 22d [wmk. 1860]

To ?; Down (type 2)
ALS; 7½ X 5; 2p.

Would be proud to be one of their series [?of photographs of famous persons], but cannot spare time for special trip to London and is not likely to be there on business soon; will call on Mr. Edwards during the summer, when next in London.

1. Watermark sets lower endpoint for date. Type of Down address employed was last used in 1861, which sets upper endpoint.

238 [1861] Feb. 2d [end. Febry 3d. 1861; pmk. FE 3/ 61]

To Ch[arles] LYELL; no location
ALS; 8 X 6½; 2p. and env., add. [Sir Ch. Lyell/ 53. Harley St/ London (W.)], end. [C. Darwin/ Febry 3d. 1861; C. Darwin/ Feb. 1861/ Agassiz &/ Bowen/ (unintelligible word--PTC)]

Printed in full, with minor changes: *Life and Letters*, II, 359-60. p. 359, line 6, change "the lengths" to "the very absurd lengths". p. 359, line 13, change "[Jean Louis Rodolphe] Agassiz admits" to "Agassiz (foolish man) admits". At end of letter is: "I sent Calcutta Review a couple of days ago."[1]

1. [Edward Blyth], "[Review of] *On the Origin of Species*," *Calcutta Review*, *35* (1860): 64-88. See also Vorzimmer, *Reprint Catalogue*, item R.61.

239 [1861] March 4th [end. March 1861; wmk. 1860]

To [Philip Lutley] SCLATER; Down (type 2)
ALS; 7 3/4 X 4 3/4; 2p., end. [C. Darwin/ March 1861]

Is working on skeletons of rabbits; wants from [Zoological] Gardens [of Zoological Society of London] one of the two Russian rabbit bucks donated by CD; will skeletonize it, and does not want skull damaged; has signed Sclater's certificate [for membership] at Royal Soc[iety of London]; would like to see a recent paper on skeleton of hybrid hare-rabbit, if Sclater has spare proof of same.[1]

1. Probably Edwards Crisp, "On Some Points Relating to the Habits and

Anatomy of the Oceanic and of the Freshwater Ducks, and also of the Hare (Lepus timidus) and of the Rabbit (L. cuniculus), in Relation to the Question of Hybridism," Proc. zool. Soc. Lond., 1861: 82-87. Read on February 26, 1861. See also Darwin, Variation under Domestication (1868), I, 126n.

240 [1861 March] 12th [end. March 1861; wmk. 1860]

To [Philip Lutley] SCLATER; Down (type 2)
ALS; 7 3/4 X 5; 2p., end. [C. Darwin/ March 1861]

Thanks for rabbit [see previous letter, above]; glad Sclater likes Asa Gray,[1] which CD incorrectly thought he himself had sent to Sclater; will soon receive corrected Darwin, Origin [(1861)], which [John] Murray will soon distribute; glad Sclater has "become 'heretical' on species"; was not surprised that Sclater was initially opposed to CD; "I cannot...respect anyone who has knowledge & can change his opinion suddenly on such a point"; please publish "a word on our side", as "those opposed write vehemently & those on our side are silent"; day before yesterday, had letter from "a Professor,[2] who dares not speak out."

1. Probably Gray, Natural Selection not Inconsistent with Natural Theology: A Free Examination of Darwin's Treatise on the Origin of Species, and of Its American Reviewers... (London: Trübner, 1861). See Life and Letters, II, 370-71. For a recent reprint of this three-part review, see Gray, Darwiniana..., ed. A. Hunter Dupree (Cambridge, Mass.: Belknap Press of Harvard Univ. Press, 1963), 72-145.
2. Perhaps George Bentham; see Life and Letters, II, 292 and 292n.

241 [1861] March 23d [end. March. 1861; wmk. 1860]

To [Philip Lutley] SCLATER; Down (type 2)
ALS; 7 3/4 X 5; 4p., end. [C. Darwin/ March. 1861/ about Birds]

Thanks for note; had not heard of Sclater's paper at Oxford;[1] could one conclude from known distribution of Gallinaceae that probability that a species of genus Gallus is endemic to South America is as low as that of endemic hummingbird from Old World; is it true that no species of Gallus is known in Africa and that probably no Gallus species wandered far from the metropolis of the genus in India and northern Malay Islands; where in Proc. zool. Soc. Lond. is described the Gallus Temminckii of G[eorge] R[obert] Gray?[2]

1. "Remarks on the Geographical Distribution of Recent Terrestrial Vertebrata," Rep. Br. Ass. Advmnt Sci., 30 (1860), pt. 2: 121-22. The B.A.A.S. met at Oxford in 1860.
2. "Notice of Two Examples of the Genus Gallus," 17 (1849): 62-63.

242 /?1861/[1] March 26th /wmk. 1860/

To ?; Down (type 2)
ALS; 7 3/4 X 5; 1p.

Thanks for corr.'s volume on old bones and for compliments.

1. Apparently the combination of Down address variant used in this letter and a watermark of 1860 is unique to the period mid-1860 to mid-1861.

243 /?1861/[1] Ap. 1 /wmk. 1860/

To ?; Down (type 2)
ALS; 7 3/4 X 5; 2p., end. /Chas. Darwin F.R.S./ Author of/ 'Origin of Species'/

Always thought corr. had many primroses; sorry for trouble; sends flowers; thanks for information about Oxalis; will repay corr. for Cypripedium and Dionaeas at one time.

1. Combination of Down address variant used with watermark provided year.

244 /1861/ Ap. 12th /end. Apr. 13/ 1861; pmk. AP 13/ 61; wmk. 1859/

To Charles LYELL; Down (type 2)
ALS; 8 X 5; 6p. and env., add. /Sir Charles Lyell/ 53. Harley St/ London (W.)/, end. /C. Darwin/ Apr. 13/ 1861/ Somme valley beds/ whether preglacial/ Ants in Texas planting/

Printed in full, with minor changes: Life and Letters, II, 364-65.

245 /1861 April/ 21. /end. April 1861; wmk. 1859/

To /Philip Lutley/ SCLATER; Down (type 2)
ALS; 7 3/4 X 5; 3p. and end. /5642/ C. Darwin/ April 1861/ on Birds of/ S. America/

Enjoyed talk with Sclater; as Sclater is working at birds of S/outh/ America, check CD's reference to three species of Opetiorhynchus in Darwin, Zoolog/y of the Voyage/ of the Beagle /(1838-1843), pt. 3/, Birds / by John Gould/, p. 67, to confirm comments on observed differences in habits of species; do similarly for Scytalopus, p. 74; has made "horrid mistake" on O/rpheus/ parvulus /pp. 63-64 and 67/, a temporary name for a form of O/petiorhynchus/ Patagonicus /p. 67/; Capt/ain C. C./ Abbot confounded O/petiorhynchus/ vulgaris and antarcticus, which CD simultaneously observed and recorded to be closely similar except in habits;[1] Opetiorhynchus /Patagonicus, pp. 67-68/ from Chiloe seems to be a case of intermediate variety.

1. Abbot, "Notes on the Birds of the Falkland Islands," Ibis, 3 (1861): 149-67; and Darwin, Zoology of the Voyage of the Beagle (1838-1843), Pt. 3: Birds, by John Gould, 66-68 and 149-50. The Abbot article is incorrectly attributed to the American, Charles Conrad Abbot, in Cat. scient. Pap., 1, 3.

246 [1861] May 4th [end. May 1861]

To [Philip Lutley] SCLATER; Down (type 4)
ALS; 8 X 5; 5p., end. [C. Darwin/ May 1861]

Thanks for note with offer to insert in Ibis a notice by CD on habitats of Falkland birds; cannot write such notice, since CD's catalogs and notes prepared on the spot refer only to specimen number, not genus and species; wrote to [George Robert] Gray and [John] Gould, but could not find original specimen, as specimens [from Beagle voyage] were given to Zoological Society [of London] and distributed; "A false habitat is a positive mischief, worse than a species not appearing in a list", so after "careful work" by Capt[ain C. C.] Abbott [sic; Abbot], better to consider the two names errors than to give them without evidence; received letter from [Robert] Swinhoe announcing delivery to Sclater of a new rock pigeon and a wild Anser cygnoides, but CD must check these claims when next in London; perhaps pigeon is Himalayan rock pigeon.[1]

1. For more on this letter, see the preceding letter above, and Darwin, Variation under Domestication (1868), I, 237.

247 [?1861][1] May 5

To [Thomas Campbell] EYTON; Down (type 4)
ALS; 8 X 5; 2p.

Has received skeletons; needs name for untagged domestic cock in longest of four boxes received; is the "'Gungla' cock" a specimen of G[allus] bankiva or G[allus] Sonneratii;[2] other boxes contain the Hamburg and the call duck; is ill.[3]

1. CD studied the osteology of fowls and ducks in May, 1861; see "Darwin's Journal," 15.
2. "Gungla" may be a derivative of the Indian (hindu) "gunga", or "market".
3. See Darwin, Variation under Domestication (1868), I, 260-70.

248 [?1861][1] May 6th

To [Thomas Campbell] EYTON; Down (type 4)
ALS; 8 X 5; 5p.

Thanks for prompt reply; without doubt, the [untagged] skeleton was duckwing game and the wild Gallus [i.e. the Gungla cock] was G[allus] bankiva, since every bone agrees; has two skeletons of Dorking, so send neither one of these nor G[allus] varius; do not send mounted skeletons; has examined 25 skeletons and about 55 skulls; skulls show only "differences characteristic of the breeds", but other bones show "much fluctuating variability"; thinks skeletons of various species of the same restricted genus or sub-genus differ only slightly; wants to quote Eyton's view that, in allied species, while there are plain differences in some parts of skeleton besides head, bones in wings and legs of all breeds are similar in configuration but not in length and thickness; will keep duck specimen [i.e. call duck; see previous letter] until CD gets to ducks in "a few weeks"; do birds with large topknot, such as curassows, have [skull] protuberances to support the topknot?[2]

1. CD worked upon fowls and ducks in May, 1861, finishing ducks on May 31; see "Darwin's Journal," 15.
2. For more on this matter, see: preceding letter; and Darwin, Variation under Domestication (1868), I, 260-70, and II, 332-33.

249 [?1861][1] May 14th

To [Thomas Campbell] EYTON; Down (type 4)
ALS; 8 X 5; 3p.

Thanks for invitation to London and to [town of] Eyton; wanted to visit London for two reasons, but is too ill; if ever visiting Shrewsbury again, will visit Eyton; answer briefly whether skeletons, except skulls, of birds of same restricted genus "do not generally very closely resemble each other", whether wing and leg bones are "generally very constant in form", and "whether in largely crested Gallinaceae the skull is protuberant to support the crest."

1. This letter clearly follows shortly after the preceding letter, above.

250 [1861] May 28th [pmk. MY28/ 61]

To James HUNT; Down (type 4)
ALS; 8 X 5, 2p. and env., add. [James Hunt Esqr/ Hon. Sec. Ethnological Socy/ 4 St. Martins Place/ London (W.C.)], end. [Darwin's/ Envelope]

Thanks to president and council of Ethnological Society for electing CD an honorary fellow; thanks for gift of first volume of Society's Transactions; thanks personally to Hunt for kind words in letter.

251 1861 June 1st

To ?; Down (type 4)
ALS; 8 X 5; 6p.

Will value information, but do not hurry with it; unclear whether corr. is thinking of "a general course of scientific experiments in crossing or only in relation to Hollyhocks"; there is open field for "research in regard to crossing varieties which have been greatly neglected under a scientific point of view, though largely & loosely practised by gardeners. Species on the other [ hand ] have been largely experimented on. As you have lived so much abroad, German is probably quite familiar to you (I wish it were to me) & I would most strongly advise you to get [Karl Friedrich von ] Gärtner['] s admirable 'Versuche ueber die Bastardzeugung. 1849'[1] & study it"; suggests in minute detail some crossing experiments with differently-colored hollyhocks which breed true; experiment outlined by CD would be "very interesting on account of a wonderful statement on this head by Gärtner with respect to crossing white & yellow Verbascum"; would suggest further experiments with Pelargonium, but must not run on.

1. Versuche und Beobachtungen über die Bastarderzeugung im Pflanzenreich (Stuttgart: K.F. Hering, 1849).

252 [1861 ] June 2d [end. June 1861 ]

To [Philip Lutley ] SCLATER; Down (type 4)
ALS; 8 X 5; 3p. and end. [5376/ C. Darwin/ June 1861/ abt Rabbits ]

Has sent two rabbits from P[orto ] Santo to the [Zoological ] Gardens [of the Zoological Society of London ] for temporary safekeeping; if they are like one brought by [ Thomas Vernon ] Wollaston in spirits, then they are curiosities, having been feral for 450 years and springing from one doe brought [to island ] by [João Gonçalvez ] Zarco; specimen CD examined differed from common rabbit in skull, shape of dorsal vertebrae, in size greatly, in coloring, in color of upper part of tail, and in ears not being edged in black; thinks this may be "a new species!!"; must get rid of rabbits because whole household leaves for a two month stay at Torquay, probably beginning the 10th, because of illness of daughter [Henrietta Emma Darwin Litchfield ]; care for rabbits, interbreed them or cross them with other rabbits, and find out what [Abraham Dee ] Bartlett [superintendent of Gardens ] thinks of them; if one or both die while CD is away, send fresh carcasses to CD in Torquay.[1]

1. For more on the rabbits, see Darwin, Variation under Domestication (1868), I, 112-15. The visit to Torquay lasted from July 1 to August 27; see "Darwin's Journal," 15.

253 1861 June 12

From Rich[ar ]d OWEN; British Museum
ALS; $7\frac{1}{4}$ X $4\frac{1}{2}$; 2p. (enclosure wanting)

Enclosed proof of note which CD will insert in his forthcoming "'Reply'" is "a correct statement of the relations of the passage I have printed on the use & meaning of the term 'Creation', as used by Naturalists in some of their discussions, to the partial quotation from it in Prof. Baden Powell's Essay."

254 [1861][1] June 15-16

To [Rev. B. S. MALDEN of Canterbury][2]; Down (type 4)
ALS; 8 X 5; 4p.

Thanks for orchids; felt "boyish delight" at *H[abenaria] viridis*, but it is not a *Habenaria*; has now seen "every British Orchid...except the Lizard [*Orchis hircina*]", which CD hopes to get from corr. or from [G. Chichester] Oxenden; will begin soon to write paper [i.e. Darwin, *Fertilisation of Orchids* (1862)]; glad to see *Aceras* specimens, but they did not have the monstrous flowers; five of six specimens from Oxenden had such flowers; old specimen of *O[rchis] fusca*, like Oxenden's specimens, showed infertility caused by infrequency of insect visits; [in left margin--PTC] look for monstrous flowers on *Aceras*; [in right margin--PTC] monstrous flowers illustrate structure of Habenaria; knows "little of Botany", but thinks unspotted purple orchids with hollow stems are *O[rchis] latifolia* (which CD once saw) and white ones are *O[rchis] maculata*; glad to see state of pollen masses on corr.'s Canterbury Fly Ophrys [i.e. *Ophry muscifera*]; look at pollen masses on Bee O[phrys, i.e. *Ophrys apifera*] and especially on its variety, [*Ophrys*] *arachnites*, to see if the masses are either removed or simply fallen on own stigmas in *oldish* flowers; return slip from *G[ardeners'] Chronicle*; "June 16th P.S.", thanks for note; part about Lizard shall be kept private; thought *O[rchis] militaris* was same as *O[rchis] fusca*.

1. This year is written in an ink similar to that used by CD, but is apparently not in CD's hand.
2. See Darwin, *Fertilisation of Orchids* (1862), 43 and 78. Malden is the only person acknowledged by Darwin to have provided specimens of the Frog Orchis (i.e. *Peristylus viridis* or *Habenaria viridis*).

255 [1861][1] June 19th Wednesday

To [?F. SMITH, of the British Museum][2]; Down (type 4)
ALS; 8 X 5; 3p.

Pollen masses attached to [specimen of] *Bombus hortorum* [i.e. common bumblebee] are not from British orchid, but are from an exotic orchid of the group Epidendreae; suspects bee was caught near a hothouse; fears that corr. does not have "one of the sand-wasps with pollen-masses attached [?which you] alluded to [*sic*]", otherwise CD would have liked to have seen it; could easily ascertain whether the [leaf?] on the S[outh] American wasp was pollen; supposes insects in corr.'s own collection do not have pollen masses attached; ask Mr. Walker ⌈?about pollen on his insects⌋.

1. Year determined from variant of Down address used, day of "Wednesday" for June 19, and information in Darwin, Fertilisation of Orchids (1862), 164.
2. Correspondent is probably either Smith or Sir W. C. Trevelyan; see Darwin, Fertilisation of Orchids (1862), 164 and in index under "Smith, Mr. F". Smith was selected because CD's mention of corr.'s own collection implies that bumblebee specimen (which was Trevelyan's) did not belong to corr.

256 ⌈?1861⌋[1] June 27th

To Dr. BULLEN; Down (type 4)
AL in third person; 8 X 5; 1p.

Thanks for sending "the orchid flowers with Diptera"; CD and a son of CD have just "made out" that "Orchis maculata is fertilised by the aid of Diptera."

1. June, 1861, was the beginning of CD's work on orchids; see "Darwin's Journal," 15.

257 ⌈1861⌋[1] July 18th ⌈wmk. 1860⌋

To ?; 2. Hesketh Crescent/ Torquay
ALS; 7 3/4 X 5; 4p.

Cannot answer corr.'s questions; began to examine living corals nearly 30 years ago, but other pursuits have interfered and CD had forgotten what he knew; had studied "the effects of tranquil & disturbed water on their growth", but forgets his conclusions on this issue; thinks nearly all species were distinct; remembers having thought that classification of stony corals would be difficult; places "much trust" in ⌈James Dwight⌋ Dana, whose health has failed, regrettably.

1. Year determined by Torquay address; see "Darwin's Journal," 15 and 15n.

258 ⌈1861⌋[1] July 20th

To ⌈Charles⌋ LYELL; 2. Hesketh Crescent/ Torquay
ALS; 7 3/4 X 5; 4p.

Printed, with minor changes: Life and Letters, II, 376. line 4, add: was "pleased, considering how many have attacked me on 'Induction' &c, to hear...from...H⌈enry⌋ Fawcett, that...J⌈ohn⌋ Mill '...considers that your ⌈i.e. CD's⌋ reasoning throughout ⌈Darwin, Origin (1859)⌋ is in the most exact accordance with the strict principles of logic.

He also says the method of investigation followed is the only one proper to such a subject.' Considering how high an authority he is, this pleases me much, & I think you will be pleased";[2] Etty [i.e. Henrietta Emma Darwin Litchfield] has improved a little. At end of letter is: is writing long paper [i.e. Darwin, Fertilisation of Orchids (1862)] on fertilization of orchids; "I almost wish I could have been completely idle here"; heaven knows when Darwin, Variation under Domestication (1868) will be done; regards to wife [Mary Elizabeth Horner Lyell].

1. Year determined by Torquay address; see "Darwin's Journal," 15.
2. See More Letters, I, 189-90 (letter 129). Cf. David L. Hull, Darwin and His Critics: The Reception of Darwin's Theory of Evolution by the Scientific Community (Cambridge, Mass.: Harvard Univ. Press, 1973), 27-28.

259 [1861 (?August 1)][1] Thursday

To [Charles] LYELL; 2. Hesketh Crescent. Torquay
ALS; 8 X 6¼; 6p.

Printed, with minor changes: More Letters, I, 190-92 (letter 130). At beginning of letter is: Emma [Wedgwood Darwin] and Etty [i.e. Henrietta Emma Darwin Litchfield] are touring, has forwarded [Mary Elizabeth Horner] Lyell's letter "with the sad account of [the death of Frances Elizabeth Appleton] Longfellow;[2] is surprised at Dutch translation [of Darwin, Origin],[3] which should be left at Q[ueen] Anne St. p. 190, line 11, change "at in my orchids is" to "at, viz. Orchids, is". p. 192, line 18, add: William [Erasmus Darwin] will join Mr. [?Edmund Gibson] Atherley's Bank, needs a good introduction to Southampton from Lyell. At end of letter is: regards to Lyell's [traveling] party.[4]

1. Year determined by Torquay address; see "Darwin's Journal," 15. Month and day presumed from date of "2 Augt 1861" written in pencil in contemporary hand (not CD's) on manuscript.
2. See DAB, XI, 383.
3. This translation is not listed in Freeman, but see Life and Letters, II, 357.
4. See Life of Lyell, II, 347.

260 [1861 August][1] 13th [wmk. 1860]

To [Charles] LYELL; 2. Hesketh Crescent/ Torquay
ALS; 8 X 5; 3p.

Printed in full, with minor changes: More Letters, I, 192-93 (letter 131).

1. Reasoning for date same as for preceding letter; date written in pencil is "August/ 1861."

261 [1861] Aug 21 [end. 1861; pmk. AU21/ 61]

To [Charles] LYELL; 2. Hesketh Crescent/ Torquay
ALS; 8 X 6½; 8p. and fragment of env., end. [C. Darwin/ returning M.S. on/ Sicily newer than species/ inhabiting it./ & on deification of Natural/ Selection./ 1861]

Printed, with minor changes: More Letters, I, 193-94 (letter 132). At beginning of letter is: returns home [to Down] Monday 26th;[1] knows page well, has quoted it; approved of note appended by Lyell; sentence only needs "trifling modification"; "adaptation of species [which allows them] to travel widely over existing continents, will necessarily adapt them for occasional still wider transportation to new lands. I have used in Origin this argument to account for very wide range of F[resh] Water productions."[2] At end of letter is: sentence at p. 3 reads roughly.

1. Actual return occurred on August 27; see "Darwin's Journal," 15.
2. See Darwin, Origin (1859), 383-88.

262 [1861][1] Sept 6th

To [Charles] LYELL; Down (type 2)
ALS; 8 X 5; 1p. (enclosure wanting)

Printed in full, with minor changes: More Letters, II, 188 (letter 524).

1. Year written on manuscript in pencil in unknown hand, but appears correct from context of letter.

263 [1861][1] Sept. 10th

To [Charles] LYELL; Down (type 4)
ALS; 8 X 5; 4p.

Thanks for interesting long letter; has no suggestions on subjects "gone out of my head"; dislikes [argument concerning] absence of organic remains"; there were [no organic remains] in Patagonia or T[ierra] del Fuego where shells were present, but from what CD has read of Greenland, suspects what Lyell now admits and [Robert] Chambers urges; is abundance of swimming animals any guide to shells, etc., living at bottom; [such] animals cannot live "where icebergs are habitually grounded"; see Darwin, ["On the Distribution of the Erratic Boulders and on the Contemporaneous Unstratified Deposits of South America,"] Trans. geol. Soc. Lond., 6 [(1842): 180-88, at] 186; sorry Lyell must "alter & modify [his published treatment of]...this great subject"; admires Lyell's industry.

1. Year written on manuscript in pencil in unknown hand, appears contemporary.

264 [1861 Sept. 15][1] Sunday Evening [?end. Sept. 15]

To [Charles] LYELL; Down (type 4)
ALS; 8 X 5; 5p., end.? [Sept. 15]

Thanks for interesting correspondence; [Thomas Francis] Jamieson is "a capital man"; has been performing experiments on Dionaea; Lyell is discussing a "grand subject", but CD cannot help with it; lake theory can account for absence of deltas on Lochaber shelves; submergence of 1,200 feet in Perthshire since glaciation is striking evidence concerning Glen Roy; has been looking at his [CD's] Glen Roy paper,[2] gives final arguments in favor of elevation and subsidence theory to explain Glen Roy; "But I suppose ice-lakes must be true cause"; disagrees with Lyell's claim in former letter that great glaciers in Scotland caused by great loftiness; glacial phenomena great in extent and prevalence [during ice age]; seems safest to assume great glacial period to be simultaneous until shown otherwise.

1. Month and day appear to be an endorsement by Lyell. Year is written in pencil on manuscript in an unknown hand and appears correct from context of letter.
2. "Observations on the Parallel Roads of Glen Roy,..." Phil. Trans. R. Soc., 129 (1839): 39-81.

265 [1861] Sept. 22d [end. Septr 1861]

To [Charles] LYELL; Down (type 2)
ALS; 8 X 6½; 5p. and end. [Darwin 81a/ Glen Roy/ Septr 1861]

Printed, with minor changes and minor omissions: More Letters, II, 188-89 (letter 525). p. 189, line 6, change "found" to "formed". p. 189, line 14, change "alluded" to "attended".

266 [1861] Oct 1st [end. 1st. Oct 1861]

To [Charles] LYELL; Down (type 4)
ALS; 8 X 5; 4p. and fragment of env., end. [C. Darwin/ answer to/ Jamieson on/ Glen Roy/ 1st. Oct 1861]

Printed in full, with minor changes: More Letters, II, 190-91 (letter 527).

267 [1861] Oct 14th [end. 1861]

To [Charles] LYELL; Down (type 4)
ALS; 8 X 5; 5p. and end. [(88)/ Darwin 1861/ on Jamieson revisit/ to Glen Roy]

Printed in full, with minor changes: More Letters, II, 191-92 (letter 528). p. 191, line 15, change "Friesland" to "Finland".

268 [1861] Oct. 20th [end. Oct. 24. 1861.]

To [Charles] LYELL; Down (type 4)
ALS; 8 X 5; 4p. and fragment of env., end. [C. Darwin/ Oct. 24. 1861./ Glen Roy glaciers/ & ice-dams/ "land straits"]

Printed, with minor changes: More Letters, II, 192 (letter 529). At end of letter is: has been working hard at orchids; "The subject is, I fear, too complex for the Public & I fear I have made a great mistake in not keeping to my first intention of sending it to Linnean Soc[iet]y; but it is now too late, & I must make the best of a bad job."[1]

1. See Darwin, Fertilisation of Orchids (1862). CD originally planned to write only a long essay on orchids, not a book.

269 [1861 October][1] 23d.

To [Charles] LYELL; Down (type 4)[2]
ALS; 8 X 5; 3p.

Has been dissecting; returns note; supposes that all believe Lyell's view that "water flowed out at head of valley where the lakes existed"; Glen Roy's "'intermediate shelf'" seemed like a plain shelf to CD, although [Thomas Francis] J[amieson] disagrees; intermediate shelf has been seen by everyone who visited Glen Roy; there is no outlet at this shelf, but [David] Milne[-Home] says there may be; valley should be searched for such outlets; "A man might spend his life there"; hopes J[amieson] will return to Glen Roy; "it is an opprobrium to British Geologists, that it shd. not be settled beyond dispute"; is disturbed by sloping, stratified, deposited detritus at all levels "by opening on a lake or arm of sea"; terminal moraine at mouth of Spean seems better than ice; "But if it were the sea, I cannot help a sneaking hope that the sea might have formed the horizontal shelves.--"[3]

1. Year and month written on manuscript in pencil in unknown hand, appears contemporary and correct, according to context.
2. Although written on stationery bearing the Down (type 4) letterhead, the page of the paper with this letterhead printed upon it is at the end of the letter; first page of manuscript text is headed simply "Down", in CD's hand.
3. See also: Darwin, "Observations on the Parallel Roads of Glen Roy,..." Phil. Trans. R. Soc., 129 (1839): 39-81; More Letters, II, 171-93; Life and Letters, I, 361-64; and Jamieson, "On the Parallel Roads of Glen Roy, and Their Place in the History of the Glacial Period," Q. Jl geol. Soc. Lond., 19 (1863): 235-59.

270 [?1861][1] Nov. 12th

To Madam [?Lady Dorothy Fanny Walpole NEVILL][2]; Down (type 4)
ALS; 8 X 5; 4p.

Dr. [John] Lindley has suggested corr. to CD as source of orchids; is preparing "small work" on orchids [i.e. Fertilisation of Orchids (1862)]; send two or three flowers of "any member of the great Tribe of Arethuseae," including "Limodoridae, Vanillidae &c.", especially "Mormodes & Cycnoches"; expects difficulty in shipping of delicate pollen masses; also wants Bonatea, Masdevillia, and "any Bolbophyllum with its lower lip or Labellum irritable"; wants "any genus with any remarkable peculiarity"; send large parcels to "'C. Darwin care of the Down Postman Bromley Kent'"; gives packing instructions.

1. This was the only November during which CD was preparing a "small work" or orchids; see "Darwin's Journal," 15.
2. Nevill was the only woman (note the "madam") acknowledged by CD in the orchid book; see Darwin, Fertilisation of Orchids (1862), 158n. The letter refers to the corr. as "your Ladyship"; Nevill was the daughter of an Earl and the wife of another relative of a peer.

271 [ca. 1861-1869][1] Feb. 24th

To ?; Down (type 4)
ALS; 8 X 5; 1p.

Thanks for proofs; cannot form any judgment, but corr.'s view is ingenious; if accepted, it will be great step in knowledge of glacier movement.

1. Years determined by Down address variant used.

272 [ca. 1861-1869][1] Ap 13

To [Peter Martin DUNCAN]; Down (type 4)
LS; 8 X 5; 3p.

C[harles] Lyell says corr. pleased to receive coral specimens from Keeling Islands; will send some via Geolog[ical] Soc[iety of London]; once had more; habitat and station for each specimen is given; collected all but one specimen himself; has a few notes about soft parts of corals.

1. Years determined by Down address variant used.

273 [ca. 1861-1869][1] Sept. 8th

To ?; Down (type 4)
ALS; 8 X 5; 3p.

Thanks for note and specimens; subject is worthy of a paper; hopes corr. has kept CD's queries; glad to hear about number of moults, but color is chief interest; tell briefly of differences of plumage of male, female, and young in two or three breeds, so CD can judge how far to pursue subject; Pile Game [a fighting fowl] would be good case; wants from breeders information of proportions of sexes of ducks and fowls; when at Manchester, find age of peacock when topknot appears; sorry corr. is ill.

1. Years determined by Down address variant used.

274 [1862 March][1]

From Cha[rles] LYELL; no location
ALS; 7 X $4\frac{1}{2}$; 4p.

Has the shingles; thanks for letter; envies CD's being almost done [with Darwin, *Fertilisation of Orchids* (1862)]; is working with printer himself; has been thinking about Glen Roy, needs CD's explanation; height of cols determines levels of [Glen Roy] shelves, not the variable heights of ice blockages; sees how ice dam caused lowest shelves in Glen Roy and Glen Spean, then another ice dam in Glen Roy raised waters even higher, but does not know how two ice blockages in one glen can cause two shelves, since lower would be destroyed when new ice blockage arrived, and since disappearance of old blockage before new blockage arrived would leave lower, formerly blocked col open for drainage; marine theory avoids this because top shelf is made first; return this note with answer, so Lyell can send it to [Thomas Francis] Jamieson; CD's brother [Erasmus Alvey Darwin] told of illness of CD's child [Henrietta Emma Darwin Litchfield].

1. This is clearly the letter which prompted the reply which follows below.

275 [1862] April 1st

To [Charles] LYELL; Down
ALS; 8 X $6\frac{1}{2}$; 3p.

Printed in full, with minor changes: *More Letters*, II, 192-93 (letter 530).

276 [1862] May 12th [end. May 1862]

To [Philip Lutley] SCLATER; Down (type 4)
ALS; 8 X 5; 3p. and end. [6105/ C. Darwin/ May 1862/ Abt. Peacock]

[Abraham Dee] Bartlett says that "Japanned Peacock" (Sclater's name for which [*Pavo nigripennis*--PTC] CD has forgotten) has appeared among ornithologist [Hudson] Gurney's birds; write to Gurney for particulars, or give Gurney's address to CD; wants to know "whether his birds appeared pure & whether any Japanned Peacocks lived anywhere near, so that there could have been a recent cross."[1]

1. See Darwin, *Variation under Domestication* (1868), I, 290-92.

277 [1862] May 14th [end. May 1862]

To [Philip Lutley] SCLATER; Down (type 4)
ALS; 8 X 5; 3p. and end. [C. Darwin/ May 1862/ Abt Birds]

Thanks for two notes; ask [Hudson] Gurney if he had any white or pied birds when P[avo] nigripennis appeared; in two of three cases mentioned by Sir R[obert] Heron [in "Notes," Proc. zool. Soc. Lond., 3 (1835): 54-55], there were whites and pieds in lot; has four cases, thinks P. nigripennis a variety, no more surprising in origin than Himalayan rabbit;[1] if [Zoological] Gardens [of Zoological Society of London] have a white and a common peacock, cross them to see if P. nigripennis appears; "the effects of crossing are sometimes marvellous in bringing out old & lost characters or in producing new characters".

1. See Darwin, Variation under Domestication (1868), I, 108-11.

278 [1862] June 15th [end. 1862]

To [George Henry Kendrick] THWAITES; Down (type 4)
ALS; 8 X 5; 4p., end. [1862]

Read Darwin, "On the Two Forms, or Dimorphic Condition, in the Species of Primula,..." J. Linn. Soc. [(Botany), 6 (1862): 77-96]; Dr. [Hugh Algernon] Weddell says cinchona presents same case of some trees with long pistils and some with short; there must be reciprocal fertilization between two forms; please check this claim on Ceylon cinchona, using artificial fertilization if necessary to cross the two forms and produce strong plants; no need to castrate; suggests this because "the growth of Cinchona is so important for mankind [as source of quinine]"; is still working at this subject; such dimorphism seems common with Rubiaceae [family of cinchona and madder]; would like analogous cases.

279 1862 June 20th

To W[illiam] B[ernhard] TEGETMEIER; Down (type 4)
ALS; 8 X 5; 3p.

Testimonial, letter of recommendation for position of curator for Hartley Institution. Excerpts printed: Life and Letters, II, 53.

280 [1862] June 20th [end. 1862]

To [George Henry Kendrick] THWAITES; Down (type 4)
ALS; 8 X 5; 4p., end. [1862]

Wrote to Thwaites on Primula [see CD to Thwaites, June 15, 1862, above] two days before receiving Thwaites's letter of May 15; is glad to hear of *Sethia*; *Menyanthes* is dimorphic, so is not surprised at *Limnanthemum*; compare by weight the output of two forms of *Limnanthemum*; on Malpighiaceae, mark the imperfect flowers, see if they set seed, see if they are closed, and see "whether the pollen-tubes are emitted from the pollen-grain within the anther & then penetrate the stigma", as is the case with imperfect flowers of *Viola* and *Oxalis*; thanks for "your Governor's letter"; "I suspect the dimorphism of *Primula*, is often, (though not at all necessarily) the high-road to dioeciousness."[1]

1. See Darwin, *Different Forms of Flowers* (1877), 116 and 122.

281 [1862] Aug. 22d. [end. Augt.24.1862; pmk. AU23/ 62]

To C[harles] LYELL; 1. Carlton Terrace/ Southampton
ALS; 8 X 5; 9p. and fragment of env., add. [Sir C. Lyell/ Albion Hotel/ Freshwater Gate/ Isle of Wight.], end. [C. Darwin./ Augt.24.1862/ Longevity of species/ in mammalia & in plants]

Printed, with minor changes: *Life and Letters*, II, 387-89. At beginning of letter is: is answering [Mary Elizabeth Horner] Lyell's letter of sympathy [for scarlet fever in family] to William [Erasmus Darwin]; Emma [Wedgwood Darwin], Lenny [i.e. Leonard Darwin], and Horace [Darwin] are still ill; will go to Bournemouth soon to be near other children; the [John William] Lubbocks have home in Chiselhurst; is glad Glen Roy is settled; moraines opposite L[och] Treig are important, as is slope inland, if proved.[1] p. 387, line 2, add: "I fancy [Thomas Henry] Huxley [i.e. *Evidence as to Man's Place in Nature* (London: Williams and Norgate, 1863)] will be out sooner"; [Joseph Dalton] Hooker says book will be interesting. p. 388, line 5, add: "What an unblushing man he must be to lecture thus after abusing me so & never to have openly retreated, or alluded to my Book [Darwin, *Origin* (1859)]." p. 388, last line, add: cannot remember reason for not adopting fully and quoting Lyell's axiom of 1832. At end of letter is: Lyell's axiom may be true; remembers "considerable perplexity on subject"; thinks mammals and molluscs are "too remote from each other for fair comparison"; is tired; regards to wife and to work.
Also printed, with minor changes: *More Letters*, I, 203 (letter 140). The *Life and Letters* version is more complete and more accurate.

1. See *Life of Lyell*, II, 358.

282 [1862] Oct. 1. [end. Sept 1. 1862]

To Ch[arles] LYELL; Down (type 2)
ALS; 8 X 5; 4p. and fragment of env., add. [Sir Ch. Lyell/ 53 Harley St/ London W.], end. [C. Darwin/ Sept 1. 1862/ Dr Falconer on Elephants/ & origin of species by/ variation/ Elephants failing/ group.]

Printed, with minor changes: Life and Letters, II, 389. At beginning of letter is: discussion of the "'Règne Humain'" is in Isid. G. [i.e. Isidore Geoffroy Saint-Hilaire], Hist. Nat. Générale [i.e. Histoire Naturelle Générale des Règnes Organiques, Principalement Etudiée chez l'Homme et les Animaux, 3v. (Paris: V. Masson, 1854-1862)], II, chap. 7, 167; is Lyell done with [Friedrich] Rolle, [?C. Darwin's Lehre von der Entstehung der Arten im Pflanzen- und Theirreich in Ihrer Anwendung auf die Schöpfungsgeschichte Dargestellt und Erläutert... (Frankfurt am Main: J. C. Hermann, 1863]?[1] At end of letter is: enjoyed chat with Lyell.

1. Apparently CD received an advance copy of Rolle's book, for he cited it with a publication date of 1862, not 1863; see Darwin, Variation under Domestication (1868), II, 308n.

283 [1862][1] December 1st

To ?; Down (type 4)
ALS; 8 X 5; 3p.

Asks three questions about cases for plants, including stove plants; each question is answered in another hand than that of CD in spaces provided in letter for answers; concerning sizes, capacities, and prices of cases.

1. This year written in pencil in what appears to be a contemporary hand at top of first page of manuscript letter.

284 [ca. 1862-1863] June 11th [wmk. 1859]

To [Henry Walter] BATES; Down (type 2)
ALS; $5\frac{1}{2}$ X $3\frac{1}{2}$; 2p. (enclosure wanting)

Answer question contained in the enclosed, written by CD's brother-in-law [Hensleigh Wedgwood] who studies and writes on language; hopes Bates's book progresses;[1] is ill.

1. The Naturalist on the River Amazons..., 2v. (London: John Murray, 1863). Years for this letter's date were determined from date of publication of this book, as well as of Hensleigh Wedgwood, On the Origin of Language (London: [Bungay], 1866).

285 [1862-1866][1] Jan. 9th

To [Thomas Campbell] EYTON; Down (type 4)
ALS; 8 X 5; 2p.

Thanks for facts about breeding; wants to be told of any other striking cases; thanks also for promise to measure webs of feet of otterhounds;

compare them with measures of foxhounds or harriers; has no opportunity to see dogs.[2]

1. Down address variant used to set lower endpoint, since this type of letterhead was introduced in mid-1861. Upper endpoint determined by date at which CD wrote last chapter of book that discusses webbed feet in dogs; see "Darwin's Journal," 17.
2. Eyton's measurements are discussed in Darwin, Variation under Domestication (1868), I, 39-40.

286 [?1863][1] Feb. 2d

To [?John Joseph BRIGGS][2]; Down (type 4)
ALS; 8 X 5; 3p.

F[rancis Trevelyan] Buckland says corr. might help CD; remembers article in Field, most likely by corr., on regrowth of fins of fish; wants citation and details of the case; wishes to quote case.[3]

1. Chapter of CD's book which refers to regeneration of dorsal fins in fish was written between Jan. 23 and April 1, 1863. See "Darwin's Journal," 16.
2. See Darwin, Variation under Domestication (1868), II, 15-16.
3. There is a two-page scrap which is catalogued with this letter; it reads: "communicated by Mr A. [Fonblanque?] of the British Consulate at Alexandria to Mr. Darwin." There is more text not worth abstracting.

287 [1863 February][1] 4th [wmk. 1860]

To [Charles] LYELL; Down
ALS; 7½ X 5; 3p.

Has just received "the great book [i.e. Lyell, The Geological Evidences of the Antiquity of Man... (London: John Murray, 1863)]"; may go to London in evening if not ill;[1] will visit Lyell some morning; has skimmed pages on species; is glad Lyell discusses many of the "most important" points "not generally touched on by others. I have read last chapt[er] with very great interest.[2] By Jove how black [Richard] Owen will look"; Lyell is too civil to Owen; "I am getting more savage against him, even than [Thomas Henry] Huxley or [Hugh] Falconer. He ought to be ostracised by every Naturalist in England"; book will "give the whole subject of change of species an enormous advance."

1. Top of first page carries in pencil, perhaps as an endorsement by Lyell, the date "4 Feby 1863". CD was in London from February 4 to February 14, 1863; see "Darwin's Journal," 16. Date of publication of Lyell's book also agrees with a date of February, 1863.
2. On contents of last chapter of the book and Lyell's feeling about it, see Life of Lyell, II, 353.

288 [1863 February] 17th [wmk. 1860]

To [Charles] LYELL; Down (type 2)
Als; 7½ X 5; 3p. (enclosure wanting)

Printed in full, with minor changes: More Letters, I, 236 (letter 162).

289 [1863] March 6th [end. March 6./ 1863; pmk. MR 6/ 63]

To C[harles] LYELL; Down
Partly ALS, partly LS; 8 X 6½; 12p. and fragment of env., add. [Sir C..../ 53..../ Lon...], end. [C. Darwin/ March 6./ 1863/ Meridional belts of warm & cold/ will not explain all the phenomena/ of distributions of species./ 8. Selection explaining adapta-/tions not enlarged enough/ More examples of rudiments./ 9. The fewness of individuals/ in islands prevents/ transmutation]

Printed, with minor changes: Life and Letters, III, 11-12. At beginning of letter is: is disappointed to have put off [a visit to] Lyell, because CD would have talked over "many points & [Richard] Owen's false letter"; Emma [Wedgwood Darwin] wants CD to stop work and visit Malvern [water cure establishment] for two months; thanks for note; keep [James Dwight] Dana, [?"On the Higher Subdivisions in the Classification of Mammals," Am. J. Sci. (Silliman's J.), 35 (1863): 65-71]. p. 12, line 14, add over seven pages of comments upon passages from [Lyell, The Geological Evidences of the Antiquity of Man...(London: John Murray, 1863)], including, inter alia, the following: thinks glacier chapters are "almost the best in the book"; closing pages of chapter 19 are "magnificent"; "I think this discussion has interested me almost more than the antiquity of man. The gloss of novelty was worn off the latter, yet I have been deeply struck by the effect of the agglomerated evidence"; "It is of little consequence, but [Joseph Dalton] Hooker published his Essay [i.e. "Introductory Essay," The Botany of the Antarctic Voyage of H. M. Discovery-Ships Erebus and Terror in the Years 1839-1843..., Pt. III: Flora Tasmaniae, 2v. (London: Lovell Reeve, 1860), I, i-cxxviii] a month after the Origin [i.e. Darwin, Origin (1859)].... I asked him"; "Who is [Nils Gabriel] Sefström?"[1] p. 12, line 16, add: wants [Richard] Owen's paper on the Aye Aye [i.e. "On the Aye-Aye,..." Trans. geol. Soc. Lond., 5 (1866): 33-101]; "I am sorely tempted to expose in Athenaeum what rubbish Owen has written on the subject."

1. Some of the other comments are paraphrased in the endorsement. CD's comments discuss the following additional persons: [John] Evans; [Bartholomew James] Sulivan; [James] Smith of Jordanhill; [John Francis Julius von] Haast; [Andrew Crombie] Ramsay; [John William] Lubbock, [Baron Avebury]; and [Christian Friedrich Hermann] von Meyer.

290 [1863] March 12th-March 13th [end. March 14 1863]

To [Charles] LYELL; Down (type 2)
ALS; 8 X 6½; 8p. and fragment of env., add. [Sir...], end. [C. Darwin/ His doctrine not so/ Lamarckian as I make it./ March 14 1863]

Printed, with minor changes: Life and Letters, III, 13-14. p. 14, line 21, add: last night Henrietta [Emma Darwin Litchfield], a "first rate critic", said: "'Is it fair that...Lyell...calls your theory a modification of [Jean Baptiste Pierre Antoine de Monet de] Lamarck's? Why is it more a modification of him, than of any one's else [sic]'"; her criticism is appropriate, but confidential; "I have more trust in your judgment than in my own, so I hope you may be right, as far as mere policy is concerned, in your very gentle statement of your belief"; thanks about Aye-Aye paper;[1] has written to "Sir Henry". p. 14, line 23, add: is interested in "Athenaeum controversy";[2] [William Henry] Flower puts well how [Richard] Owen "has falsely dragged in size of brain"; Lyell's answer good, except too civil in saying that Owen "must have forgotten what he said in Annals;[3] it was a brazen lie, & ought not, I think, to have been treated so delicately." At end of letter is: has reread Lyell's letter; is ill; starts for [hydropathy at] Malvern after Easter holidays."[4]

1. See preceding letter, above.
2. See Life and Letters, III, 8, 8n, and 9-10.
3. "On the Cerebral Characters of Man and the Ape," Ann. Mag. nat. Hist., 7 (1861): 456-58.
4. CD visited Malvern from September 2 to October 14, 1863; see "Darwin's Journal," 16.

291 [1863] March 17th [end. March.17.1863; pmk. 63]

To C[harles] LYELL; Down (type 2)
ALS; 8 X 6½; 8p. and fragment of env., add. [Sir C. Lyell/ 53. Harley St/ London W.], end. [C. Darwin/ on Lamarck--on other au-/thors claims for notice in/ 'Antiquity of Man'/ March.17.1863]

Printed, with minor changes: Life and Letters, III, 15-17. p. 16, line 2, add: is astounded and grieved at part of Lyell's letter to [Joseph Dalton] Hooker about conduct of [Hugh] Falconer in "the monkey-case"; early part [of Lyell, The Geological Evidences of the Antiquity of Man,... (London: John Murray, 1863)] had too many references to [Lyell,] Principles [of Geology..., 3v. (London: John Murray, 1830-1833)]; is weak. p. 17, line 1, change "forgotten)," to "forgotten, Count Laperda or Saperde or some such name [i.e. Louis Charles Joseph Gaston, Marquis de Saporta]),". p. 17, line 8, add: [John William] Lubbock [Baron Avebury] is not satisfied with [Lyell's] notice of his Somme paper ["On the Evidence of the Antiquity of Man, Afforded by the Physical Structure of the Somme Valley," Nat. Hist. Rev., 2 (1862): 244-69].[1] p. 17, line 10, add: Hooker forgets "that he told me himself [the] date of publication of his Essay."[2] p. 17, line 12, add: [Hooker] has more plants from Cameroon Mountains and will discuss mundane cold period. p. 17, line 13, missing element is "[Richard] Owen's [paper on the] Aye-Aye".[3] p. 17, line 17, add: "& I had written so good a letter (!) all ready, with a blank for his [i.e. Owen's] sentence claiming more than he had any right to; but I could pick out no such sentence. Hooker says he so despises him that he cannot hate him: I do not know whether this [is] a right frame of mind, but by Jove it is not my frame of mind." At end of letter is: agrees about [Thomas Henry] Huxley and "the Review [i.e. Natural History Review: A Quarterly Journal of Biological Science]", which is excellent nevertheless;[4] as CD has spoken against entomologists, see notice by CD on [Henry Walter] Bates's paper ["Contributions to an Insect Fauna of the Amazon Valley," Trans. Linn. Soc. Lond., 23 (1862): 495-566] on mimetic resemblances.[5]

1. The notice is probably Lyell, The Geological Evidences of the Antiquity of Man... (London: John Murray, 1863), 19n.
2. See CD to Lyell, March 6, [1863], above.
3. See: ibid.; and preceding letter, above.
4. See Life of Lyell, II, 366.
5. See Nat. Hist. Rev., 3 (1863): 219-24.

292 [1863][1] March 23

To [William Darwin] FOX; Down (type 4)
AL (mutilated); 8 X 5; 3p.

Glad to hear authentic particulars and Fox's own case; mistakenly read 23 for 13 lambs and supposed that breeding was over two seasons; shows preponderance of one color over other; has exzema on face, so may postpone trip to Malvern.

1. Letter to Fox dated March 9, 1863, and preserved at Christ's College, Cambridge, mentions the birth of 23 lambs and trip to Malvern. Another letter to Fox dated May 23, 1863, preserved in same place, mentions that eczema is better. Particulars in this letter are in Darwin, Variation under Domestication (1868), II, 30-31; this was written in early 1863, according to "Darwin's Journal," 16.

293 [1863][1] March 30th

To [George Henry Kendrick] THWAITES; Down (type 4)
ALS; 8 X 5; 4p.

Thanks for specimens; Sethia is a "beautiful case" of reciprocal dimorphism; could not compare pollen because long-styled anthers were knocked off; long stamens dust the bodies of insects visiting these flowers, while short stamens dust proboscis; unequal lengths of alternate stamens insures that pollen on the proboscis is from more than a pair of anthers, as seen by CD in Lythrum; Ceylonese Lythraceae, if any, will show three forms, with two kinds of pollen in each flower; wants more facts and specimens; thanks for Limnanthemum, a "very pretty case";[2] has hothouse for growing it, but needs seeds; [ in margin--PTC ] flower with different color in each half anther, if not due to abortion, indicates dimorphism or trimorphism; wants to experiment on all orders; presumes that Sethia is a tree; will send soon a "little paper" on dimorphism of Linum [ "On the Existence of Two Forms, and on Their Reciprocal Sexual Relation, in Several Species of the Genus Linum," J. Linn. Soc. (Botany), 7 (1864): 69-83 ]; thanks for fact concerning Discospermum;[3] thanks for specimens of galls, which interest CD, who is ignorant of them; almost experimented once upon galls; "It is truly wonderful what a change a little poison or irritation has effected, & effected in so diversified a manner";[4] thinks [ ?galls on ] specimens of Gomphia and Lesemia result from insect puncture or disease, not from sports or bud variation.[5]

1. Year written on top of first page of original in what appears to be a contemporary hand--perhaps an endorsement.
2. See Darwin, Different Forms of Flowers (1877), 116, 122, and 249. See also CD to Thwaites, June 20, [ 1862 ], above.
3. See Darwin, op. cit., 286.
4. See Darwin, Variation under Domestication (1868), II, 282.
5. CD here refers to Lesemia coccinea, now known as Salvia microphylla (Kunth). These generic names were identified by Professor Joseph Ewan, Tulane University.

294 [1863][1] Ap. 18th [wmk. 1860]

To [Charles] LYELL; Down (type 2)
ALS; 4p. @ 7 3/4 X 5, 2p. @ 8 X 5; 6p.

Printed, with minor changes: Life and Letters, III, 20-21. At end of letter is: invites Lyells to Down; has had eczema; book by [Henry Walter] Bates [i.e. The Naturalist on the River Amazons..., 2v. (London: John Murray, 1863)] is "capital".

1. Year written, perhaps in endorsement, at top of first page of original. Year of publication of book by Bates supports this year.

295 [1863][1] July 29th

To [George Henry Kendrick] THWAITES; Down (type 4)
ALS; 8 X 5; 4p.

Thanks for Limnanthemum seeds;[1] knows about conjugation of algae from pamphlet sent years ago by Thwaites which records a double embryo in Fuchsia to which CD has recently referred;[2] thanks for specimen of and information on Cassia, which has curious flower; have [Rev. S. O.] Glenie see how insects visit it and if pistil moves during inflorescence; can not send carte de visite, since none has ever been made of CD; sends instead a "poor small photograph" taken by CD's son [?Leonard Darwin].

1. Year written on top of first page of original in what appears to be a contemporary hand--perhaps an endorsement. For CD's request for the seeds, see CD to Thwaites, March 30, [1863], above.
2. See: Thwaites, "On Conjugation in the Diatomaceae," Ann. Mag. nat. Hist., 20 (1847): 9-11, 343-44; and idem, "Further Observations on the Diatomaceae; with Descriptions of New Genera and Species," ibid., 1 (1848): 161-72. CD possessed and quoted from the latter article; see: Vorzimmer, Reprint Catalogue, item 25; and Darwin, Variation under Domestication (1868), I, 391 and 391n.

296 [1863][1] Aug. 14th

To [Charles] LYELL; Down (type 4)
ALS; 8 X 5; 4p.

Printed, with minor changes: More Letters, I, 243 (letter 171). At beginning of letter is: thanks for letter; congratulations on finding Arctic shells;[1] [Edward Bissell] Hunt sent copy of his paper ["On the Origin, Growth, Substructure and Chronology of the Florida Reef," Am. J. Sci. (Silliman's J.), 35 (1863): 197-210];[2] was struck by it, but [James Dwight] Dana has reservations; "poor Dana" must restrict his mental exertions; glad that Antiquity [i.e. Lyell, The Geological Evidences of the Antiquity of Man... (London: John Murray, 1863)] sells well, "but if it gets another edition of 'Origin" [i.e. Darwin, Origin] I shall not be grateful; for I dread the very thought of that job." line 9, add: CD's sister [Emily] Catherine [Darwin Langton] will marry [Charles] Langton [on October 8, 1863], who had been married to [Charlotte Wedgwood Langton], sister of Emma [Wedgwood Darwin]; a good match.[3] At end of letter is: regards to wife; "I answered the Manchester Pigeon man for you."

1. Year written in pencil on top of first page of original in what appears to be a contemporary hand--perhaps an endorsement. On arctic shells, found in 1863, see Life of Lyell, II, 379-80.
2. See Vorzimmer, Reprint Catalogue, item 290.
3. Cf. Emma Darwin, II, 202-03.

297 [1863][1] Dec 29

To [George Henry Kendrick] THWAITES; Down Bromley Kent, S.E.[2]
Copy of L; 7¼ X 4½; 4p.

Wants cases of bud variations, called sports by gardeners, such as "moss-rose in Provence", in which leaf bud assumes new character; asks this because Sir R[obert Hermann] Schomburgk says flowers introduced to St. Domingo [i.e. Santo Domingo, Dominican Republic] from warm temperate regions often had such bud variations;[3] wrote "some time ago" about dimorphic cinchonas,[4] and is now more certain of this, with some plants "absolutely sterile" with their own form of pollen; compare size of pollen of two Ceylonese genera mentioned by Thwaites, or send dried specimens; is "hard at work" on Darwin, <u>Variation under Domestication</u> [1868].

1. Year written on top of first page of original in an apparently contemporary hand. See also note 4, below.
2. This appears to be a handwritten facsimile of Down (type 4).
3. See Darwin, <u>Variation under Domestication</u> (1868), I, 373-87 and 401-11, esp. 408.
4. See CD to Thwaites, June 15, [1862], above.

298 [?1864][1] Ap 7

To ?; Down (type 4)
LS; 7 X 4½; 4p.

Corr. will see tubes if he places pollen on stigma, places pistil under strong microscope twelve to eighteen hours later, and tears up stigma; once one knows appearance of tubes with plant like geranium, one never mistakes them; has been in bed for six months, so cannot exchange orchids; Catasetums hard to exchange; Messrs [James] Veitch [& Sons] said they were not valuable enough to be marketable; some bi-generic crosses have been made, but seeds do not germinate.

1. Down address variant used limits endpoints to 1861 and 1869. According to "Darwin's Journal," 15-18, esp. 16, CD's only six-month illness during this period was from October 14, 1863, to April 13, 1864.

299 1864 Ap 18.

To [David J. BROWN][1]; Down (type 4)
LS; 7 X 4½; 4p.

Thanks for letter and MS; is ill and must be brief; corr.'s paper is "striking & original", but "all the Journals have so often discussed the 'Origin of Species' that I do not believe any of them wd insert any other paper on the subject"; honors men like corr. who spend "much time in labour" and yet find time to "acquire extended knowledge & follow out original trains of thought"; returns MS.

1. This letter is part of the David J. Brown Papers, which were acquired in a single lot.

300 [1864][1] June 28th [wmk. 1860]

To Fanny [i.e. Frances Mackintosh WEDGWOOD]; Down
ALS; 8 X 5; 3p.

Sends condolences ⌈on death of her son, James Mackintosh Wedgwood⌋; remembers corr.'s kindnesses at Malvern with ⌈death of⌋ "poor Annie" ⌈i.e. Anne Elizabeth Darwin⌋; regards to ⌈husband⌋ Hensleigh; write back when ⌈daughter⌋ Hope ⌈Elizabeth Wedgwood⌋ writes to ⌈CD's daughter⌋ Etty ⌈i.e. Henrietta Emma Darwin Litchfield⌋.

1. James Mackintosh Wedgwood died in 1864; see Emma Darwin, I, xxvii.

301 ⌈?1864⌋[1] July 30

To ⌈William⌋ BOWMAN; Down (type 4)
LS; 8 X 5; 2p.

Thanks for note; will attend to information when CD publishes book ⌈?Expression of the Emotions (1872)⌋; thanks for kindness to son; illness delayed writing.

1. Down address variant used determines endpoints of 1861 and 1869. Of CD's two long illnesses during this period, only that from October 14, 1863, to April 13, 1864, would provide CD with excuse for not writing until July.

302 ⌈1864 September 6⌋ ⌈end. Sept.6.1864.--/ Sept. 28th. 64.⌋

To Asa GRAY; no location
Address leaf only; 8 X 6½; 2p., add. ⌈Prof Asa Gray/ Cambridge/ Massachusetts/ U. States⌋, end. ⌈Handwriting of/ Charles Darwin, Scit./ Sept.6.1864.--/ Sept. 28th. 64.⌋

303 ⌈?1864⌋[1] Dec 8

To ⌈?Charles Victor NAUDIN⌋[1]; Down (type 4)
LS; 8 X 5; 3p. (enclosure wanting)

Thanks for letter, photograph, and congratulations; encloses the only photograph of CD available, one taken by son ⌈?Leonard Darwin⌋; thanks for reference to "the Comptes Rendus" on subject of special interest about which CD has few authentic facts; glad corr. is still working on Cucurbitaceae; recently, has quoted extensively from corr.'s papers on this "order ⌈sic; family⌋" in work being prepared on variation;[1] will send soon a paper on Lythrum ⌈Darwin, "On the Sexual Relations of the Three Forms of Lythrum salicaria," J. Linn. Soc. (Botany), 8 (1865): 169-96⌋.

1. See Darwin, Variation under Domestication (1868), I, 357-60, for extensive quotation of Naudin on Cucurbitaceae. This chapter was written in late 1864; see "Darwin's Journal," 16-17.

304 [1865] Jan 22

To [Charles] LYELL; Down (type 4)
LS; 8 X 5; 8p.

Printed in full, with minor changes: Life and Letters, III, 32-34. p. 34, line 2 after signature, change "address." to "Address & tell him about Sexual Selection."

305 [?1865][1] Jan 27.

To [John Edward] GRAY; Down (type 4)
LS; 8 X 5; 1p.

Thanks for congratulations;[1] sorry that corr. is ill; corr. looked "far from well" when CD last saw him.

1. The congratulations were probably for CD's Copley Medal of the Royal Society of London, awarded November 30, 1864; see "Darwin's Journal," 16n. Perhaps the illness of Gray was the cause of Gray's belatedness in congratulating CD.

306 [1865] Feb 21

To [Charles] LYELL; Down (type 4)
LS; 8 X 5; 4p.

Printed in full, with minor changes: Life and Letters, III, 35-36. At end of letter is: regards to [Mary Elizabeth Horner] Lyell from [CD's wife and amanuensis for this letter, Emma Wedgwood Darwin].[1]

1. This final remark is in Emma Darwin's hand and is signed with her initials.

307 [1865] March 25th [wmk. 1860]

To [Charles] LYELL; Down (type 4)
ALS; 4p. @ 8 X 5, 1p. @ 7 3/4 X 5; 5p. (enclosure wanting)

Thanks for enclosed correspondence; ought CD to identify Miss Buckley as source about roosting in trees--the only new point--or would "received through Sir C. Lyell" be adequate; pigeons in upper Egypt settle in flocks on low trees, but not on palms; "the duke [?i.e. George Douglas Campbell, Duke of Argyll] making such a point as this rests on the Lamarckian belief that everything in structure & habits must change;[1] I have put the case that such a change, if not selected or induced by compulsion, would be a downright difficulty on my notions"; more curious case is that of domestic pigeons which resemble a flock of gulls when they settle upon and float down the Nile to drink where the banks are perpendicular; "I have read most of H[erbert] Spencer's Biology [i.e. The Principles of Biology, 2v. (London: Williams & Norgate, 1864-1867)] & agree with you. Some of his remarks are very clever & suggestive, but somehow I seldom feel any wiser after reading him, but often feel mistified [sic]"; Spencer dictates, so he has "detestable" style; [Joseph Dalton] Hooker thinks Spencer's last number is best ever written; has finished Elements [i.e. Lyell, Elements of Geology, 6th ed. (London: John Murray, 1865)], was struck by Lyell's "summing up on the Laurentian stages"; health still poor, but works two hours per day on "my 'Domesticated Animals & Cultivated Plants [i.e. Variation under Domestication (1868)]'."

1. See Campbell, "Opening Address," Proc. R. Soc. Edinb., 5 (1866): 264-92, esp. 281. See also Life and Letters, III, 31-34.

308 [1865] May 6 [end. 1865; pmk. MY 6/ 65]

To J[ames] P[hilip] Mansel WEALE; Down (type 4)
LS; 8 X 5; 4p. and env., add. [J. P. Mansel Weale Esq/ (crossed out--PTC) Port Elizabeth/ (crossed out--PTC) Algoa Bay/ Cape of Good Hope/ (added in margin--PTC) 4/ (added in margin--PTC) Adelaide], end. [First letter from C.D./ 1865 in answer to one/ from (P.L.?) 1864]

Thanks for tracings and letter of February 19; is ill; case of mule is interesting, but CD has heard of similar; the only aids in naming Cape orchids are "extremely expensive illustrated works"; cannot suggest any good zoological works either; "Few foreign countries are so well off botanically as the Cape will be with [William Henry] Harvey's work [viz. Harvey and Otto Wilhelm Sonder, Flora Capensis: Being a Systematic Description of the Plants of the Cape Colony, Caffraria, & Port Natal, 3v. (Dublin: Hodges, Smith, and Co., 1859-1865)]"; do not study fertilization of orchids, since enough has been written on it, but see Rob[er]t Brown, ["On the Organs and Mode of Fecundation in Orchideae and Asclepiadeae"], Trans. Linn. Soc. Lond., [16 (1833): 685-745], as there is a good deal to be learned on fertilization of this genus; has seen an American hymenopter with pollen masses of Asclepias [milkweed] covering its tarsi, but does not know how pollen is placed upon stigma; research on caverns of S[outh] Africa will be interesting; must bring or send specimens to Europe in order to name them.

309 [1865][1] June 4

To ?; Down (type 4)
LS; 6 3/4 X 4½; 2p.

Thanks for note; is "much better", but is "far from strong" and is "living in my bedroom"; was stupid about the books; 1843 volume is right; have ready by Thursday morning the volume for 1844 containing on p. 295 a paper by /̅René Joachim Henri_7 Dutrochet /̅i.e. "Recherches sur la Volubilité des Tiges de Certains Végétaux et sur la Cause de ce Phénomène," C. r. hebd. Séanc. Acad. Sci., Paris, 19 (1844): 295-303_7; will then return the wrong volume.

1. Date determined by Sydney Smith, St. Catharine's College, Cambridge, based upon state of CD's health; see "Darwin's Journal," 16-17.

310 /̅1865_7[1] June 24th

To ?; Down (type 4)
ALS; 8 X 5; 3p.

Thanks for copy of corr.'s "new work on the 'Races of the Old World'"; thinks view on correlation of color and constitution, expressed in remarks on p. 388, is probable; a year ago, through director general of medical department of British Army, CD circulated printed questions which asked about correlation of tropical diseases with color of hair and skin of victims to all regimental surgeons in the tropics; doubts that responses will be reliable.

1. CD sent out the questionnaire mentioned here in May, 1864; see Life and Letters, III, 90.

311 /̅1866_7 Jan 6 /̅end. Jan. 1866_7

To /̅Philip Lutley_7 SCLATER; Down (type 4)
LS; 8 X 5; 3p. and end. /̅8361/ C. Darwin/ about China Ducks/ Jan. 1866_7[1]

/̅Robert_7 Swinhoe writes to CD of domestic Chinese duck perhaps descended from Anas poeciloryncha /̅spot-billed duck from India_7; is this a species distinct from A. boschas /̅common wild duck, or mallard_7; what are differences; does it have curled tail feathers and wing marks, and is it a native of China; "After nine months inaction from illness, I have just begun to do a little work".

1. Alongside endorsement, there are some penciled notes by Sclater, mostly illegible, but with "not at all like A. boschas" very clear.

312 /̅1866_7 Feb 7. /̅wmk. 1860_7

To /̅Charles_7 LYELL; Down (black border)[1]
LS (postscript in CD's hand); 7 X 4½; 10p.

Printed in full, with minor changes: More Letters, I, 476-78 (letter 363). p. 477, line 26, add: CD and Emma /̅Wedgwood Darwin_7 send love to Lady /̅Mary Elizabeth Horner_7 Lyell. p. 478, line 2, change "S. on Himalaya" to "S. & on Himalaya".

1. CD's sister, Emily Catherine Darwin Langton, died on February 2, 1866; see "Darwin's Journal," 17.

313 [1866] Feb. 15th--Thursday

To [Charles] LYELL; Down (black border)[1]
LS; 7 X 4½; 6p.

Printed in full, with minor changes: *More Letters*, I, 478-79 (letter 364).

1. See preceding letter, above.

314 [1866][1] Feb. 22

To [Charles] LYELL; Down.
ALS; 7 X 4½; 3p.

Glad [Charles James Fox] Bunbury has pointed out CD's errors; probably has notes about *Drimys*, *Fuchsia*, etc. in "my Portfolio on this subject"; seems to remember that "some of fossil mammals of Caves of Brazil are Andean, as Vicunas, Bears & Goat-like animal &c."; is especially glad to have seen these letters because CD heard this morning that [John] Murray wants a new [fourth or 1866] edition of *Origin* & CD "must alter a few words about Organ Mountains"; might add that [Jean Louis Rodolphe] Agassiz has detected glacial markings on these mountains;[1] is sad that "my present work [i.e. writing of *Variation under Domestication* (1868)]" will be stopped for one to three months.[2]

1. Year determined by year of publication of edition of Darwin, *Origin* containing change about Organ Mountains. See Darwin, *Origin* (1866), 444-45. Cf. *ibid.*(1861), 405. Or see Peckham, *Variorum Origin*, 595.
2. For dates of interruption of "my present work", see "Darwin's Journal," 17.

315 [1866 March 3][1] Saturday

To [Charles] LYELL; Down (black border)[2]
ALS; 7 X 4½; 4p.

Has returned memorial, hopes it is successful; on mundane cool period, could have given additional facts if CD had known of Lyell's interest; worked in new facts in last German edition of *Origin* [?i.e. *Freeman* 428] which will appear in new [fourth] English edition [i.e. *Freeman* 115], but will be too late for Lyell [to include in the first volume of the tenth edition of Lyell, *Principles of Geology*, published November 1866]; one important consideration is that "it can be proved that individuals of the same plant, growing N. & S, or growing on mountains & plains, certainly become acclimatised & transmit different constitutional powers of withstanding cold to their seedlings; & this would come into play with the slowly advancing glacial period";[3] CD's MS. of 47 folios, written ten years ago, is available if Lyell wants it; has read "a paper on the representative closely allied Petrels of N. & S. Oceans.--"

1. A date of "5 March 1866" is written on the first page in pencil. Context of letter obviously follows the preceding letter above and precedes the subsequent letter below. March 5, 1866, was a Monday.
2. CD's sister, Emily Catherine Darwin Langton, died on February 2; see "Darwin's Journal," 17.
3. See Darwin, Origin (1866), 448.

316 [1866] Mar 8

To [Charles] LYELL; Down/ Bromley Kent (handwritten) (black border)
LS; $7\frac{1}{4}$ X $4\frac{1}{2}$; 4p.

Printed, with minor changes: More Letters, II, 157-58 (letter 506). At end of letter is: sends MS., thinks most useful pages are those tied in green ribbon; these passages appeared in second German and French editions [of Darwin, Origin (1862-63 and 1866, respectively)] and will appear in the next [i.e. fourth] English edition, due this summer; return these special pages in a week; rest of MS. may be kept for longer period; it is the old MS. abstracted for [Darwin,] Origin; sends also "some pencil notes & a letter from [Joseph Dalton] Hooker after he had read this 10 yr old M.S."[1]

1. The old MS. was a part of Stauffer, ed., CD's Nat. Selection. The pencil notes may be the passage reproduced in More Letters, I, 438. Hooker's letter is printed in More Letters, I, 437-38 (letter 333). For Lyell's reply, see Life of Lyell, II, 408-09. For the passage under discussion, see Darwin, Origin (1866), 442ff; or see Peckham, Variorum Origin, 591-610.

317 [1866 late spring][1] Monday

To ?; Down (type 4)
ALS; 8 X 5; 3p.

Thanks for letter with facts on birds admiring themselves; new [fourth] edition of Darwin, Origin contains remarks on beauty, but does not go into details; will send corr. a copy of it when published in the summer.

1. The extended discussion on beauty was added to the fourth edition, which CD finished on May 10, 1866. See "Darwin's Journal," 17.

318 [?1866][1] Aug 2d

To ?; Down (type 4)
ALS; 8 X 5; 3p.

Thanks for "remembering what interests me"; has not seen "Bäer's [sic; Baer's] paper," but has read "long extracts in one of [Rudolph] Wagner's papers after publishing my historical sketch [in 1860-1861]";[2] knew about case of hairy and toothless family (so like the Turkish dog) through [John] Crawfurd's [Journal of an] Embassy [from the Governor-General of India to the Court of Ava, with an appendix...by Professor Buckland and Mr. Clift, 2nd ed., 2v. (London: Henry Colburn, 1834)] and through "[Jules?] Travels"; did not know case was brought before B.A.A.S.; is working on "the very obscure subject of the causes of variability in domestic production" and hopes to go to press in winter or spring.[3]

1. Year determined by date of printing of Darwin, Variation under Domestication (1868). See "Darwin's Journal," 17.
2. Wagner's paper is actually a book, Zoologisch-Anthropologische Untersuchungen (Gottingen: n.p., 1861). See Darwin, Origin (1866), xx-xxi; or see Peckham, Variorum Origin, 69, line 68.1:d.
3. CD here refers to Variation under Domestication (1868); see note 1.

319 [1866] Sep 8 [and] Sunday Morning [Sept. 9]

To [Charles] LYELL; Down (type 4)
LS (postscripts in CD's hand); 4p. @ 8 X 5, 5p. @ 7 X 4½; 9p.

Printed, with minor changes: More Letters, II, 159-60 (letter 507). At beginning of letter is: postponement of Lyell's visit has been great disappointment; CD's sister [Susan Elizabeth Darwin] suffers greatly and there is no hope of recovery;[1] is glad that Lyell may come [to Down] in October.

1. She died on October 3; see "Darwin's Journal," 17.

320 [1866] Oct 9th

To [Charles] LYELL; Down. (black border)[1]
ALS; 7¼ X 4½; 4p.

Printed in full, with minor changes: More Letters, I, 271-72 (letter 192).

1. See note to preceding letter, above.

321 [1866] Oct 12

To [Charles] LYELL; Down (black border)[1]
LS (postscripts in CD's hand); 4p. @ 7¼ X 4½, 1p. @ 7 3/4 X 5; 5p.

Has read sheets [of the MS. of the amended chapters of Lyell, Principles of Geology, 10th ed., 2v. (London: John Murray, 1866-1868)] with "enthusiastic admiration"; returns them; thinks this the best thing published by Lyell; has no serious criticisms, but gives three remarks on survival of cold period by mammoth (p. 188), on permanence of continents (slip, p. 14), and on absence of secondary and plutonic rock on islands of great oceans (slip 15); gives congratulations for near-completion of Lyell's work; passage about evaporation of snow, which is conjectural but correctly reported, is in Darwin, Journal of Researches [(1839), 277-78n, or (1845),] 245n; [Charles] Pritchard, in his Nottingham sermon,[2] says that, according to [George Biddell] Airy, [John Couch] Adams, and others, the day is slowly increasing in length, so that a trillion years ago its length was 1/200 second, and it will be 80 years long a trillion years hence.

1. See note to second letter preceding, above.
2. The Continuity of the Schemes of Nature and of Revelation: A Sermon [on Eccl. iii. 14, 15] (London: 1866). This sermon was preached at the meeting of the B.A.A.S. in Nottingham in 1866; see DNB, 46, 404.

322 [1866] Dec. 1

To [Charles] LYELL; Down. (black border)
ALS; $7\frac{1}{4}$ X $4\frac{1}{2}$; 3p.

Emma [Wedgwood Darwin] has brought Lyell's "grand book";[1] thanks for it and for inscription on first page; Lyell's additions are prodigious; wants to read them, especially climate chapter, but must read five or six books and long papers first; "for p. 136, read 146"; congratulations.

1. Principles of Geology, 10th ed., 2v. (London: John Murray, 1866-1868), I.

323 [1866][1] Dec. 1st

To Mrs. [Anne] MARSH-CALDWELL; Down Bromley Kent (black border)
ALS; $7\frac{1}{4}$ X $4\frac{1}{2}$; 4p. (enclosure wanting)

Thanks for note; returned yesterday from stay in London with Erasmus [Alvey Darwin], who is not very well; does not know Christian name or address of Mr. Corbet, so please forward enclosed note to him; has given Corbet "all the information I could", but "mere chance experiments in diet" are useless; Emma [Wedgwood Darwin] is ill.

1. Date determined by Sydney Smith, St. Catharine's College, Cambridge University, who also states that Corbet was a blind friend of Mrs. Marsh-Caldwell, whose daughter Rosamund stayed with Corbet.

324 [1866] Dec 24 [end. Dec 1866]

To [Philip Lutley] SCLATER; Down Bromley/ Kent. (handwritten)
LS; 7 3/4 X 5; 1p. and end. [9402/ C Darwin/ Enclosure/ Dec 1866]
and 2p. enclosure

Sends enclosed so Sclater can "communicate directly with the man" if desired. Enclosure is a portion of an undated ALS from Benjamin Dann Walsh, presumably addressed to CD, which is entirely crossed out except for the following: "If you happen to know any Ornithologist who wishes to exchange European Birds for North American birds, I have a particular friend here, 'Dr. Velie, Rock Island, Illinois,' who has a very fine collection & is an excellent manipulator of birds'-skins. He is also an honorable man to exchange with, which is more than can be said of certain naturalists. But don't put yourself to any trouble on account of this matter."[1]

1. Jacob W. Velie was Curator of the Chicago Academy of Sciences from 1879 to 1893; see A. T. Andreas, History of Chicago from the Earliest Period to the Present Time, 3v. (Chicago: A. T. Andreas Co., 1884-86), III, 430-31. Suzanne W. Brown, Secretary to the Director, Chicago Academy of Sciences, provided this information.

325 [?1867] Jan 31.

To [George Henry Kendrick] THWAITES; Down Bromley Kent (handwritten)
LS; 8¼ X 5½; 2p. (enclosures wanting)

J[ames] Emerson Tennant [sic; Tennent] says captured elephants weep when moaning and screaming; observe this, and see if the "orbicularis palpebrarum" acts to wrinkle the surrounding skin and partially or wholly close the eyes; encloses "some printed copies of my queries on expression, with two of the more important ones a little amended"; would appreciate "a few observations on any race".[1]

1. On the subjects discussed in this letter, see: Darwin, Expression of the Emotions (1872), 21, 167-68, and 167n; and R. B. Freeman and P. J. Gautrey, "Charles Darwin's Queries about Expression," Bull. Br. Mus. nat. Hist. (hist. Ser.), 4 (1972): 205-19.

326 [1867] Feb 22. [pmk. FE 22/ 67]

To J[ames] P[hilip] Mansel WEALE; Down (type 4)
LS; 8 X 5; 4p. and env., add. [J. P. Mansel Weale Esq/ Bedford/ Algoa Bay/ Cape Colony], end. [Feb: 1867 in answer to a/ letter from Bedford]

Thanks for letter and paper on Bonatea,[1] which CD read and forwarded to Linnean Society; several points in it are new to CD; as it has published on orchids before and is in arrears, the Society may not publish the paper, but it is a valuable paper nonetheless; Weale's observations on Asclepias are new and curious; Robert Brown said years ago that he did not understand how pollen masses were retained by stigmas which do not emit viscid matter; has not heard of Brown's conduct as botanist in Weale's colony; thanks for praise of CD's work; "It is a most serious drawback to me that I am very seldom able to go to London or to see any of my fellow-workers in Natural History ewing to my constant state of ill-health".

1. Weale, "Notes on the Structure and Fertilization of the Genus Bonatea, with a Special Description of a Species Found at Bedford, South Africa," J. Linn. Soc. (Bot.), 10 (1869): 470-76.

327 [1867][1] May 7.

To [John Maurice] HERBERT; Down (type 4)
LS; 8 X 5; 3p.

Thanks for letter; knew [Richard] Dawes as an undergraduate, but had seen him only once since; will send two guineas for his memorial; remembers that [Marmaduke] Ramsay of Jesus [College, Cambridge] told Dawes he would do nothing but laugh through life, and all agreed, which shows "how little a man some times knows himself"; is "much better" in health after "two or three bad years", but [John] Tyndall gives "too flourishing an account", as CD is "never well for the whole day"; declines invitation; invites the Herberts to Down House.

1. Dawes died on March 10, 1867.

328 [1867] June 1st

To [Charles] LYELL; Down (type 4)
ALS; 8 X 5; 4p. (enclosure wanting)

Printed, with minor changes: Life and Letters, III, 65-66. At beginning of letter is: does not think H[enry] Parker ever reviewed the Origin; perhaps Lyell refers to an article[1] on [George Douglas Campbell,] D[uke] of Argyll, which CD praised and a copy of which CD encloses; please return it.

1. Probably "The Reign of Law," Saturday Review, 23 (1867): 82-84. But see also: Saturday Review for November 15, 1862; and Life and Letters, III, 274.

329 [1867] June 9th

To [Charles] LYELL; Down (type 4)
ALS; 8 X 5; 3p.

The "variability & passage of Primrose into Cowslip must be given up", since CD has just proved that common oxlip is "a natural Hybrid between the two", but "Bardfield oxlip, which occurs almost only in Essex (the

P. Elatior of Jacquin) is a perfectly distinct & good & third species"; is glad Lyell likes H/ enry 7 Parker's article;[1] wants to discuss N/ orth 7 British Review and to see the Lyells; should go to London on fifteenth and, "stomacho volente", will breakfast with Lyell on Monday the seventeenth;[2] Lyell's note about Primula lacked enclosure.

1. See preceding letter, above.
2. CD was in London June 17-24, 1867; see "Darwin's Journal," 17.

330 /1867 7[1] June 10

To / Charles 7 KINGSLEY; Down (type 4)
LS; 8 X 5; 9p. (incomplete) (enclosures wanting)

Thanks for deeply interesting letter; lends paper by Cap/ tain Frederick Wollaston 7 Hutton, "a very acute observer", from a collection of reviews of Darwin, Origin (1859+);[2] has just finished / George Douglas Campbell, 7 Duke / of Argyll 7's book / The Reign of Law (London: Alexander Strahan, 1867) 7 and / Fleeming Jenkins, "The Origin of Species," 7 N. Br. Rev., /46 (1867): 149-71 (issued June, 1867) 7; Duke's book "very well written, very interesting, honest & clever & very arrogant. How cooly / sic 7 he says that even J/ ohn 7 S/ tuart 7 Mill does not know what he means"; parts of book are weak, such as about rudimentary organs and structure of humming birds; criticizes Duke's argument regarding diversity of structure; see CD's comments on this in Darwin, Origin (1866), 226, and on beauty, p. 238; sends copy of Origin to Kingsley; see also enclosed letter by / Alfred Russel 7 Wallace; Duke speaks absurdly of beauty "existing independently of any sentient being to appreciate it"; thinks deity should not come into scientific discussion; glad Kingsley admits sexual selection, which CD has studied recently; on lack of beauty in female birds, Wallace finds relation between nature of nest and female's beauty;[3] female peacock merely selects beauty, not every detail of color, so circular spot could become circular zone; "correlation of growth" is bad term, now prefers "correlation of variation"; Duke's attack on this is unfair; Duke wrong about natural selection; "I presume he wd not deny that / Robert 7 Bakewell, Collins, &c had in one sense made our improved breeds of cattle, yet of course the initial variations have naturally arisen; but until selected, they remained unimportant, & in this same sense natural selection seems to me all-important"; N. Br. Rev. is telling and hostile, but lacking in knowledge; reviewer wrong that domestic races formed rapidly; regarding antiquity of world, "I cannot implicitly believe the mathematicians, seeing what widely different results / Samuel 7 Haughton /, William 7 Hopkins & / William 7 Thompson / sic; Thomson, Lord Kelvin 7 have arrived at";[4] note from / Charles 7 Lyell, just received, dismisses N. Br. Rev.; study of geological work done during glacial period impresses one with necessary lapse of time; who wrote N. Br. Rev.?

1. CD read the Duke's book during this year; see Life and Letters, III, 65.
2. CD probably refers to Hutton's review of the Origin in the

Geologist (1861): 132-36 and 183-88. See Life and Letters, II, 362; More Letters, II, 183-84; and Vorzimmer, Reprint Catalogue, item R.79.
3. See More Letters, I, 283 (letter 203); ibid., II, 59-61 (letter 429); and ibid., II, 72-74 (letter 440).
4. See Joe D. Burchfield, "Darwin and the Dilemma of Geological Time," Isis, 65 (1974): 301-21; and idem, Lord Kelvin and the Age of the Earth (New York: Science History Publications, 1975).

331 [1867] July 18

To [Charles] LYELL; Down (type 4)
LS; 8 X 5; 6p. and sketch

First portion printed, with minor changes: Life and Letters, III, 71-72. At end of this portion is: thanks about six-fingered men, but that chapter [i.e. chapter 12 of Darwin, Variation under Domestication (1868)] is finished.
Next portion printed, with minor changes: More Letters, II, 141-43 (letter 493). At end of letter is: [Ludwig] Rütimeyer sent his book [Über die Herkunft Unserer Thierwelt: Eine Zoogeographische Skizza... (Basel: H. Georg, 1867)], but has not read it or cut pages; wishes Lyell good progress in book [The Principles of Geology, 10th ed., 2v. (London: John Murray, 1866-1868), volume 2].

332 [1867] Aug 22

To [Charles] LYELL; Down (type 4)
LS (postscript in CD's hand); 8 X 5; 4p.

Printed, with minor changes and minor omissions: Life and Letters, III, 72-73. p. 73, line 2, missing phrase is "with all its bad judgment & taste". At end of letter is: regards to wife [Mary Elizabeth Horner Lyell]. At beginning of postscript is: leave French edition [of Darwin, Origin (Freeman 420) (1866)] at 6 Queen Anne St. [home of CD's brother, Erasmus Alvey Darwin]. p. 73, line 2 of postscript, change "Russian has" to "Russian, Kowelowsky [i.e. Vladimir Onufrievich Kovalevsky], has".

333 [1867] Aug 24 [end. 24th. August 1867]

To [?Cassell, Petter, & Galpin, publishers][1]; Down (type 4)
LS; 8 X 5; 3p. and end. [923/ C. Darwin/ 24th. August 1867]

Has had much correspondence with [Vladimir Onufrievich] Kovalevsky, who has visited Down House; he is brother of "distinguished naturalist" [Aleksandr Onufrievich Kovalevsky]; he is preparing a translation of Darwin, [Variation under Domestication (1868)],[2] and has translated some expensive German works; he has made liberal offers to CD for translation; trusts Kovalevsky, but cannot answer for his "pecuniary circumstances", which are probably good.

1. This correspondent was suggested in a private correspondence to the author from Professor James A. Rogers, Department of History, Claremont Men's College.
2. For citation of this translation in Russian, which is impossible to reproduce here, see B. Mus. Catalogue, 48, 964, second entry from top.

334 [1867] Oct 4

To [Charles] LYELL; Down (type 4)
ALS (portion not in CD's hand); 8 X 5; 4p.

Printed in full, with minor changes: More Letters, I, 299-300 (letter 217).

335 [1867] Oct 26

To [George Henry Kendrick] THWAITES; Down (type 2)
LS; 5½ X 4½; 1p. and enc.[1]

Since corr. has helped CD with queries, encloses a few slightly corrected copies; stir up any likely and accurate man.

1. Enclosure is one-page broadside, Queries about Expression, identical to that identified as version number 2 in Richard Broke Freeman and Peter Jack Gautrey, "Charles Darwin's Queries about Expression," Bull. Br. Mus. nat. Hist. (hist. Ser.), 4 (1972): 207ff.

336 [1867] Oct 31st

To [Charles] LYELL; Down (type 4)
LS (postscript in CD's hand); 8 X 5; 4p.

Printed, with minor changes and minor omissions: More Letters, II, 4-5 (letter 381). At beginning of letter is: Emma [Wedgwood Darwin] has headache. p. 4, line 10, change "distances" to "differences [sic]". At end of letter is: brother [Erasmus Alvey Darwin] is home now.

337 [1867] Dec 7

To [Charles] LYELL; Down (type 4)
LS; 8 X 5; 3p.

Printed in full: More Letters, I, 284-85 (letter 205).

338 [1867] Decr. 9th [end. Dec 1867]

To [Philip Lutley] SCLATER; Down (type 4)
ALS; 8 X 5; 2p. and end. [10392/ C. Darwin/ abt. Index &c/ Dec 1867]

Wants both indices, since has no index of Proc. zool. Soc. Lond.; send to same address as "Journal &c"; wants more recent list of fellows than 1858 version which CD has at present; December number of Intellectual Observer does not contain article by Sclater; supposes article on barbets is in November number;[1] enjoys talk with Sclater.

1. Supposition is correct. See Sclater, "Barbets, and Their Distribution," Intellectual Observer, 12 (1867-1868): 241-46. Talk with Sclater probably occurred while CD was in London in late November and early December; see "Darwin's Journal," 17.

339 [?1867][1] Dec. 21st

To [Albert Charles Lewis Gotthilf GÜNTHER]; Down (type 4)
ALS; 8 X 5; 1p., end. [to A Günther 1870]

Thanks for note; information is just what CD wanted; has looked at plates and text; cases are "capital & quite new to me"; should have known that corr.'s work would have this in it.[2]

1. Apparently, this letter is a reply to a letter from Günther dated 19 December 1867. This letter is bound in volume 82 of the Darwin Papers at University Library, Cambridge. See Handlist of Darwin Papers, 21.
2. CD probably refers to Günther, Catalogue of the Fishes in the Collection of the British Museum, 8v. (London: 1859-1870).

340 [?1867][1]

To [?Julius Victor CARUS][1]; no location
ALS (incomplete); 10½ X 8; 1p.

Gives definitions of the following terms: shoulder-pad; cross and intercross; structure; constitution. Shows differences in the terms "constitution", "habit", and "instinct". Is astonished and pleased that translation[1] will be ready by May.

1. The translation is probably the first German edition of Darwin, Variation under Domestication [Freeman 459] (1868). This information, the year for this letter, and the recipient were suggested by Sydney Smith, St. Catharine's College, Cambridge University.

341 [?1867-1868][1] Feb. 13th

To [George Henry Kendrick] THWAITES; Down (type 4)
ALS; 8 X 5; 3p.

Printed, with minor changes and minor omissions: More Letters, II, 90-91 (letter 455).

1. See: Darwin to Thwaites, Jan. 31, [?1867], above; "Darwin's Journal," 17-18; and Darwin, *Descent of Man* (1871), II, 284ff and 319.

342 [?1867-1868][1] May 19th

To [George Henry Kendrick] THWAITES; Down (type 4)
ALS; 8 X 5; 4p.

Thanks for note of April 1; case of monkeys is hopeless; take no more trouble about human expression; learned after writing that orbicularis palpebrarum contracts on trumpeting elephants in Zoological Gardens, and distress brings tears; would appreciate further details; [Edgar Leopold] Layard says a breed of fowls in Ceylon has hen alone colored like sooty white plumage; are young of breed sooty like hen or pure white like cock?

1. For further discussion of points in this letter, see the following letters above: Darwin to Thwaites, Jan. 31, [?1867]; *ibid*., Oct. 26, [1867]; and *ibid*., Feb. 13, [?1867-1868]. See also Darwin, *Variation under Domestication* (1868), I, 256 and 256n.

343 [?1867-1868][1] Sept. 28th

To [George Henry Kendrick] THWAITES; Down (type 4)
ALS; 8 X 5; 2p.

Thanks to Thwaites and to [Edgar Leopold] Layard for trouble; is surprised at result, since Layard is trustworthy and [Edward] Blyth says same about fowls in Bengal; do not send fowls, since Zoological Society foolishly objects to domestic varieties.

1. See preceding letter, above.

344 [1868][1] Jan 30

To [John Maurice HERBERT]; Down (type 4)
LS; 8 X 5; 3p.

Thanks for congratulations [because son George Howard Darwin was second wrangler at Cambridge]; son has pleased CD; reminds CD of celebration of [Charles Thomas] Whitley's wrangler honors and honors of corr.; old microscope, received anonymously [from Herbert], reminds CD of their old friendship; regards to wife.

1. See *Emma Darwin*, II, 216-17. This letter is part of the Herbert collection.

345 [1868] Feb. 28 [end. March 1868]

To [Philip Lutley] SCLATER; Down (type 4)
ALS; 8 X 5; 1p. and end. [200/ C. Darwin/ to Secretary/ March 1868]

Thinks Capt. [Philip Parker] King's birds [?collected on 1826-1830 voyage of H.M.S. Adventure] were given first to Zoological Society and then to British Museum, but habitats for many specimens are incorrectly marked.

346 [1868 (?March 3-10)][1]

To [Charles] LYELL; 6 Queen Anne St.
ALS; 8 X 5; 4p.

Thanks for long letter, somewhat consoling, "but I take [the argument for a short age of the earth based on a calculation of the combustion rate of] the Sun much to heart"; is consoled by argument based on ignorance of universe and of process causing incandescence in heavenly bodies; formed opinion that continents have had long endurance on basis of facts of geographical distribution; see Darwin, Origin [(1959), chap. 11, esp. 357-58] and Darwin, Coral Reefs; agrees about denudation, had concluded that matter brought to lower level was much ground up; this would not affect "Scotch streams"; is tired from morning in British Museum.

1. Year is penciled on letter by CD's son. During 1868, CD was at 6 Queen Anne Street only during March 3-10 and November 7-16; see "Darwin's Journal," 17-18. March period is preferred because of date on which William Thomson, Lord Kelvin, read his famous paper on the age of the earth; see Thomson, "On Geological Time," [read February 27, 1868], Trans. geol. Soc. Glasg., 3 (1871): 1-28.

347 [?1868] Mar 4

To [? Thomas RIVERS]; Down (type 4)
LS; 8 X 5; 3p.

Thanks for note with details on Le Compte family and on moss roses [Rosa centifolia]; regrets having missed recent papers by corr. on inheritance, "but from living in the country & not seeing periodicals I have no doubt missed much of importance"; hopes CD has acknowledged corr. adequately.

348 [1868 ca. March 6] [end. March/ 1868; pmk. MR6/ 68; wmk. 1860]

To J[ohn] Jenner WEIR; 6 Queen Street/ Cavendish Sqe/ W.
ALS; 8 X 5; 7p. and env., add. [J. Jenner Weir/ 6. Haddo Villas/ Blackheath/ S.E.], end. [Darwin/ March/ 1868]

Printed in full, with minor changes: More Letters, II, 66-68 (letter 435).

349 [1868 March 19][1] Thursday

To [Charles] LYELL; 4 Chester Place/ N.W
ALS; 8 X 5; 3p.

Thanks for "the grand Book [i.e. Lyell, Principles of Geology, 10th ed., 2v. (London: John Murray, 1866-1868), volume 2]"; has read only the newspaper since coming to London [on March 3], and vowed not to look at Lyell's book, but curiosity forced him to read some of the organic part; Lyell has given a "fair history of the progress of opinion on Species; but you will perhaps think it would be strange if I did not say so, seeing what high credit you give to me"; is delighted that Lyell alludes to pangenesis; "an untried hypothesis is always dangerous ground"; "My fear has always been that Pangenesis would be a still-born infant, over whom no one would rejoice or cry"; is eager to read chapters on domestication and on man.

1. Full date written on first page in pencil.

350 [1868] May 22nd. [pmk. MY 23/ 68]

To D. PENNETHORNE; Down (type 4)
LS; 8 X 5; 2p. and env., add. [D. Pennethorne Esq/ 1 New Square/ Lincolns Inn/ London]

Thanks for note and paper on man;[1] has studied the question for years and is writing essay on it, so found nothing new in corr.'s paper; agrees with corr. on "almost all points".

1. See Vorzimmer, Reprint Catalogue, item Q. 131. CD's essay is probably his book, Descent of Man (1871).

351 1868 June 8

To "Gentlemen" [?Bibliographisches Institute in Hildburghausen, Germany]; Down (type 4)
LS; 8 X 5; 4p.

Has suggested to [John] Murray the publication of an English edition of [Alfred Edmund] Brehm's [Illustrirtes] Thierleben [6v. (Hildburghausen: Bibliographisches Institute, 1864-1869)], but Murray thinks England saturated with [T.W.] Wood, [?Curiosities of Ornithology]; thinks Brehm's book is "quite excellent, & the illustrations are admirable"; wants to borrow stereotypes of some of Brehm's illustrations for Darwin, [Descent of Man (1871)]; thinks this will further advertise Brehm's work; will pay for stereotypes; wants from volume 1, the three monkeys pp. 54, 57, and 119; from volume 3, bower bird p. 317, Paradisea p. 324, Cosmetornis p. 669, Rupicola p. 745, and Cephalopterus p. 752; from volume 4, hummingbird pp. 109 and 111, [Crepidonia?] p. 326, Polyplectron p. 471, Machetes p. 625, Palamedea p. 739, and Rhynchaea p. 743.[1]

1. For stereotypes used by CD, see Darwin, *Descent of Man* (1871), II, 42, 47, 59, 70, 75, 88, 90, 202, 283, 309, and 311. Publisher of Brehm's work was assumed to be recipient of this letter.

352 1868 July 14

To [Charles] LYELL; Down (type 4)
LS; 8 X 5; 4p.

Printed, with minor changes: *More Letters*, II, 160-61 (letter 508). At beginning of letter is: thanks for [Constantin Wilhelm Lambert Glöger [sic; Gloger], [?*Das Abandern der Vogel...* (Breslau: 1833)[1]], but already has it, so has sent it to Zoological Society; thanks also about hairy men, which CD possesses; [George Henry] Lewes has published three or four articles altogether, but not consecutively and more are yet to appear;[2] differs from Lewes on many points, but finds independent thinking on subject refreshing.

1. See Darwin, *Variation under Domestication* (1868), II, 298.
2. See *Cat. scient. Pap.*, *3*, 995; *8*, 216; and *12*, 445.

353 [?1868] Aug. 25th

To [Thomas Campbell] EYTON; Down (type 4)
ALS; 8 X 5; 2p.

Glad to see Eyton's handwriting; thanks for Eyton's Osteology [i.e. *Osteologia Avium: A Sketch of the Osteology of Birds* (Wellington: R. Hobson, 1867)]; congratulates Eyton on perseverance; had heard of Eyton's skeleton of Opisthocomus; sorry about Eyton's eyes; was too ill to have been at [B.A.A.S. mmeting at] Norwich [August 19-26, 1868]; remembers having hunted and fished at Eyton as a youth.

354 1868 Sep 2.

To [George Henry Kendrick] THWAITES; Down (type 4)
LS; 8 X 5; 2p.

Thanks for "taking all my questions so very kindly"; Thwaites has entered on fowl question in earnest; thanks to [S. O.] Glenie for "excellent letter"; has written a letter of thanks in return.[1]

1. See Darwin, *Expression of the Emotions* (1872), 21, 167, and 252. See also letter from Darwin to Thwaites, May 19, [?1867-1868], above.

355 1868 Sept. 25

To ?; Down (type 4)
ALS; 8 X 5; 1p.

Has received specimen and forwarded it to [Joseph Dalton] Hooker; will write when he hears Hooker's verdict; congratulates corr. for zeal.

356 1868 Oct 25 [pmk. OC 25/ 68]

To Carl Ritter von SCHERZER; Down (type 4)
LS; 8 X 5; 2p. and env., add. [Ministerial Rath/ Dr Carl Ritter von Scherzer/ Alsergrund/ Liechtensteinstrasse/ Vienna/ via Belgium/ No 2.]

Thanks for letter of October 20 and offer of aid; thanks for copy of the instructions; thanks for "translating & inserting the questions on expression";[1] good wishes for voyage [to East Asia].[2]

1. See Darwin, Expression of the Emotions (1872), 15-16.
2. See Poggendorff, III, 1184.

357 [1868] Dec. 12th

To [William Darwin] FOX; Down (type 4)
ALS; 8 X 5; 1p.

Thanks for letter just received; will find Fox's return on the sheep and cattle very useful;[1] must prepare a new [fifth] edition of "that everlasting Origin";[2] "I am sick of correcting."

1. See Darwin, Descent of Man (1871), I, 304-05.
2. This edition was prepared between December 26, 1868, and February 10, 1869; see "Darwin's Journal," 18. This is basis for determination of year for this letter.

358 1868 Dec 29.

To [George Henry Kondrick] THWAITES; Down (type 4)
LS; 8 X 5; 4p.

Thanks to Thwaites and friends for taking trouble about fowls; results doubtful if not the pure breed; is "unwilling to believe that [Edward] Blyth & [Edgar Leopold] Layard were both mistaken"; do males of black chickens mentioned by Layard undergo change when they become adult; is glad to see [S. O.] Glenie's letter; thank Glenie and Layard; is astonished that Glenie saw no contraction of orbicular muscle in elephants, since CD and others have seen this plainly on elephants in Zoological Gardens; perhaps too great a contraction is expected; thanks for specimen showing surprising difference in form of leaves.[1]

1. See letter to Thwaites, September 2, 1868, above.

359 1869 Jan 24

To [Thomas Campbell] EYTON; Down (type 4)
LS; 8 X 5; 3p.

Cannot provide needed information; glad Eyton continues to work on osteology of birds; has not seen Alph[onse] M[ilne] Edwards's book [probably Recherches Anatomiques et Paléontologiques pour Servir à l'Histoire des Oiseaux Fossiles de la France, 4v. (Paris: V. Masson et Fils, 1867-1871), II, published in 1868 or 1869], but hears it is excellent; how many months after birth do horns appear in males of fallow deer?[1]

1. See Darwin, Descent of Man (1871), I, 288n.

360 1869 Jan 27

To [John] PHILLIPS; Down (type 4)
LS; 8 X 5; 2p.

Thanks for book on Vesuvius [Vesuvius: A History of the Mountain and of Its Successive Eruptions (Oxford: 1869)], which has not yet arrived.

361 1869 Jany 31

To [James CROLL]; Down (type 4)
LS; 8 X 5; 6p.

Printed in full, with minor changes: More Letters, II, 162-64 (letter 510).

362 1869 March. 3rd.

To [Thomas Roscoe Rede STEBBING]; Down (type 4), but with "S.E." crossed out in hand of CD's amanuensis
LS; 8 X 5; 2p.

Printed in full, with minor changes: Life and Letters, III, 110-111.

363 [1869] March 4th

To ?; Down (type 5)
ALS; 8 X 5; 3p.

Heard while recently in London that corr. had "already started" [on a journey of some sort]; gave spare copy of corr.'s book to [Philip Lutley] Sclater to forward to Dr. Hartlant, the poor ornithologist; do camels, when uttering a loud and prolonged sound, contract their eyelids and skin around eyes; cannot observe this at Zoological Gardens, because camels there make no noises.

364 [1869] March 5th

To [Charles] LYELL; Down (type 4)
ALS; 8 X 5; 4p.

Lyell's point is good; when reading [James] Croll,[1] thought of preservation of glaciation in Scotland and Wales and of turf-covered shelves of Glen Roy and elsewhere; supposed Croll would say wear was confined to hard rocks in valley; but according to Lyell's figures, water channels could not supply enough detritus; someone should visit Glen Roy to measure area of naked rock and detritus; thinks shelves unchanged since they bordered the water; thinks Croll's and [Archibald] Geikie's estimate of surface degradation[2] is too high, even if estimate of present rate is correct; "Somehow I cannot persuade myself that the Glacial period was much more recent than hitherto supposed"; would like to hear Croll's reply to this; sorry about "Amazonian shells".

1. "On Geological Time, and the Probable Date of the Glacial and the Upper Miocene Period," Lond. Edinb. Dubl. Phil. Mag., 35 (1868): 363-84; and 36 (1868): 141-54, 362-86.
2. Ibid.; and Geikie, "On Modern Denudation," Geol. Mag., 5 (1868): 249-54, reprinted in Trans. geol. Soc. Glasg., 3 (1871): 153-90.

365 1869 March 5th.

From Thomas R[oscoe] R[ede] STEBBING; Tor Crest Hall/ Torquay.
Copy of L, retained by author; 7 X 4½; 1p.

Was prejudiced against CD's theories when he began to read CD's works, but CD's writings convinced him of truth of CD's theories; thanks for "kind letter" approving Stebbing's paper on Darwinism.[1]

1. See letter to Stebbing, March 3, 1869, above.

366 [?1869][1] March 6th

To [Thomas Campbell] EYTON; Down (type 4)
ALS; 8 X 5; 1p.

Received fawn's head safely, but previous information sufficed; pleased to hear that CD's son at Trinity [?Francis Darwin, George Howard Darwin, or Horace Darwin] is a friend of Eyton's son [?William Campbell Eyton, Thomas Slaney Eyton, or Robert Henry Eyton]; "It is a resurrection of old times [when CD and Eyton were at Cambridge together]."

1. This letter appears to follow letter to Eyton dated January 24, 1869, above.

367 [1869] March 20th

To [Charles] LYELL; Down (type 4)
ALS; 8 X 5; 4p.

Thanks for [Alfred Russel] Wallace's letter;[1] had not formed opinion on [Henry] Moseley's or [James] Croll's papers,[2] but latter seemed ingenious; since Moseley studied mechanics, he must have thought of Wallace's considerations; thought parallel veins in glacial ice and curve formed in straight row of sticks [arranged transversely in ice across top of glacier] show that glaciers do not descend by succession of abrupt fractures; subject is "beyond me"; Wallace's book [The Malay Archipelago..., 2v. (London: Macmillan, 1869)] is capital, especially discussions on geographical distribution, and it will be a success; is working hard.

1. See letters from Wallace to Lyell dated March 13 and March 17, 1869, Lyell Papers, APS Library.
2. Moseley, "On the Mechanical Possibility of the Descent of Glaciers by Their Weight Only," Proc. R. Soc., 17 (1869): 202-08; and Croll, "On the Physical Cause of the Motion of Glaciers," Phil. Mag., 37 (1869): 201-06.

368 1869 March 29

From [Jean Louis] A[rmand] de QUATREFAGES [de Bréau]; Paris
?ALS; 8½ X 6¼; 4p. (in French)[1]

Wanted to meet--and tried to meet--CD's son, but without success; hopes to meet CD some day; thanks for compliments on studies of Darwinism; although he opposes CD, Quatrefages acknowledges CD as only man to propose an evolutionary theory which is scientific and embraces all problems posed by organic nature; thinks scientific reason is still on Quatrefages's side, but appreciates grandeur of CD's work and says so in final installment, to appear April 7; will send a copy of this paper to CD; through printer's error, incorrectly stated in third article of series that CD erred in discussion of single origin for titmouse and nutcracker; hopes their differences of opinion never alter their good relations.

1. Translation into English provided by editor's wife, Nancy E. Carroll.

369 [1869] May 4th

To [Charles] LYELL; Down (type 5)
ALS; 8 X 5; 4p.

Printed, with minor changes: Life and Letters, III, 116-17. p. 117, line 15, missing phrase is "& weak". At end of letter is: has recovered from crushing by horse in three weeks instead of in three months as [James] Paget expected.[1]

1. See Emma Darwin, II, 226.

370 1869 May 20

To [Charles] LYELL; Down (type 5), but "Beckenham" has not been written in
LS; 8 X 5; 4p.

Printed in full, with minor changes: More Letters, II, 144-45 (letter 495).

371 [?1869][1] May 21st

To [William Winwood READE]; Down (type 4)
ALS; 4p. @ 8 X 5, 2p. @ 7 X 4½; 6p. (enclosure wanting)

Thanks for letter; has long been interested in expression; enclosed queries sent to various parts of world; has received only a few answers, chiefly from Australia; none discussed "true negroes"; thus, would be grateful for answers; corr. will find observation difficult; gives suggestions on how to facilitate observation; has had no answers on query five, "an expression well known to the old Grecian statuaries";[2] was just wishing for observer on Guinea coast;[1] do horns and ruffs of throat hair appear first on breeds of sheep in which rams alone acquire such traits or on breeds in which both sexes acquire them; give age when these traits first appear; on gorilla and chimpanzee, has contradictory evidence on whether upper or lower body of gorilla is hairiest; presumes voice of male the more powerful; do "the wild natives" consider "their own characteristic features" to be beautiful; do barbarian women influence particular men to woo them or to purchase them?[3]

1. Reade went to the Guinea coast in 1869; see DNB.
2. See: Darwin, Expression of the Emotions (1872), 21-22, 279, and 289; and Richard Broke Freeman and Peter Jack Gautrey, "Charles Darwin's Queries about Expression," Bull. Br. Mus. nat. Hist. (hist. Ser.), 4 (1972): 205-19.
3. See Darwin, Descent of Man (1871), I, 289n, and II, 247, 285, 344, 346, 350, 357, and 374.

372 [1869] June 30th

To [Edgar Albert SMITH][1]; Caerdeon/ Barmouth/ N. Wales
ALS; 8 X 5; 4p.

Belated thanks for proofs of excellent woodcuts; make female in figure 2a less dark; give name of male; return proofs, MS. list of specimens or instructions for woodcuts, and revise of corrected proof previously submitted by CD; may need more cuts; is ill.

1. Written in pencil on last page is: "To E. A. Smith/ Zool. Dept./ British Museum".

373 [ca. 1869] Aug. 4th

To Miss LLOYD; Down (type 5)
ALS; 8 X 5; 1p. and env., add. [Miss Lloyd/ Caerdeon] (enclosure wanting)

Thought corr. would like to see enclosed letter from [William] B[oyd] Dawkins, who was pleased by visit, although it was not as geologically interesting as expected; "There is something wonderfully curious in coming across remains of our old savage progenitors, with their ground-down teeth."

374 [ca. 1869-1871][1] April 18th

To ?; Down (type 5)
ALS; 8 X 5; 4p.

Would like to come to London to be photographed, but cannot spare time; has already refused about six such requests, "but none with a quarter of the regret that this causes me"; invites corr. to come to Down via "Orpington station (on the S.E.R. Charing Cross)" to take photo; corr. could easily reach sixteen good men without troubling about CD.

1. Years determined by Down address variant used.

375 [ca. 1869-1871][1] April 23d

To [St. George Jackson MIVART]; Down (type 5)
ALS; 8 X 5; 2p.

Thanks for "promptly answering my note"; had taken inaccurate notes on corr.'s opinions; hopes corr. will continue labors on primates; "Whenever I publish my book [?Descent of Man (1871)] I can see that I shall meet with universal disapprobation, if not execution.-- The truth is hard to gain, however much one may try."

1. Years determined by Down address variant used.

376 [ca. 1869-1871][1] Aug 24th/25th

To ?; Down (type 5)
ALS; 8 X 5; 3p.

Thanks for copies of three publications; will send duplicates to Linnean Society; cannot read Italian, unfortunately, but wife will translate parts of corr.'s works; is glad corr. discussed pangenesis, even if corr. is critical of it; will have this discussion translated carefully, will read it, and will submit it for publication; P.S. has just received highly complimentary letter from corr. dated August 22.

1. Years determined by Down address variant used. Corr. *could* be F. Delpino, but this is nothing more than a guess.

377 [ca. 1869-1881][1] July 20th

To [Jean Louis Armand de QUATREFAGES de Bréau]; Down (type 7) (pink paper)
ALS; $8\frac{1}{4}$ X $5\frac{1}{2}$; 2p.

A German friend has asked for complete list of CD's publications, so sends copy of list to corr., though fuller than required; thanks for "the interest...taken about my election [to the French Académie des Sciences]"; number of votes for CD, published in newspapers and owed to efforts of corr., was "far higher than I had expected."

1. CD began to use Beckenham in his address in 1869 and he died in April, 1882. He was elected a correspondent of the botanical section of the French Academy on August 5, 1878, after years of ballotting.

378 1870 Feb. 5th

To the Secretary [of the American Philosophical Society, Charles B. TREGO]; Down (type 5)
ALS; 8 X 5; 1p., end. [*Read March 4, 1870*/ *accepts* ]

Thanks for letter of October 15, informing CD of election to membership in APS; accepts with thanks.

379 [?1870][1] May 28th

To [Jean Louis Armand de QUATREFAGES de Bréau]; Down (type 7) (pink paper)
LS; $8\frac{1}{4}$ X $5\frac{1}{4}$; 5p.

Thanks for volume containing "wonderfully clear & able discussion" and fair treatment of CD;[1] repeated mention of CD is gratifying; first two parts are "overly supportive" of CD; in subsequent parts, strictures are severe but fair; agrees about some criticisms; does not think *Parus* and nuthatch (or *Sitta*) are directly related, as French translation [?of Darwin, *Variation under Domestication* (1868)] implies; *Canis Magellanicus* is too foxlike in appearance and voice to be a feral domestic dog; "It is curious how nationality influences opinion: a week hardly

Figure 6. Charles Robert Darwin, probably in his late 50's or early 60's. This photograph, a Darwin *carte de visite* of the period, was taken by the London Stereoscopic and Photographic Companv. (American Philosophical Society)

passes without my hearing of some naturalist in Germany who supports my views, & often puts an exaggerated value on my works; whilst in France I have not heard of a single zoologist except M. [A.] Gaudry (and he only partially) who supports my views"; hopes to influence French "embryo naturalists"; thanks for lectures;[2] thanks for kindness to son George [i.e. George Howard Darwin] in Paris.

1. Quatrefages, Charles Darwin et ses Précurseurs Français... (Paris: Baillière, 1870). Year for letter determined by this publication date.
2. See Vorzimmer, Reprint Catalogue, item Q.134.

380 1870 July 30 [pmk. JY 30/ 70]

To J[ames] P[hilip] Mansel WEALE; Down (type 5)
LS; 8 X 5; 4p. and env., add. [J. P. Mansel Weale Esq/ Brooklyn/ near King William's Town/ Kaffraria], end. [Letter on Disperis &c collected/ on Kagaberg Bedford/ 1868/70]

Thanks for letter of May 25 and for MS., drawings, and specimens; sent latter three to Linnean Society, but there will be no meeting until November; was most struck by what Weale says about parallelism of Disa and Ophrys, about structure and three nectaries of Disperis, about contraction of caudicles, about mares and asses, about acrid secretion from the Soldier-termes, and about case of bee's nest; Linnean is backlogged and impoverished, so they may not publish papers; papers could be condensed; few are interested in fertilization of flowers; sorry to hear bad account of Weale's colony.[1]

1. See: Weale, "Notes on a Species of Disperis Found on the Kagaberg, South Africa," J. Linn. Soc. (Bot.), 13 (1870-1872): 42-45; idem, "Some Observations on the Fertilization of Disa macrantha," ibid., 45-47; idem, "Notes on Some Species of Habenaria Found in South Africa [abstract]," ibid., 47-48; and idem, "Observations on the Mode in Which Certain Species of Asclepiadeae are Fertilized [abstract]," ibid., 48-58. All papers communicated by CD and read on November 3, 1870. See also: [Weale], "Natural History Notes from South Africa," Nature, Lond., 10 (1874): 486-87; and Darwin, Descent of Man (1871), I, 416-17.

381 [1870][1] July 31st

To [David] FORBES; Down (type 5)
ALS; 8 X 5; 3p.

Thanks for proof sheets (to page 272) of Forbes's paper on the Aymara;[1] has had to alter little; will refer readers to Forbes rather than giving details;[2] on page 221, ten lines from bottom, for "longer" read "shorter".

1. Forbes, "On the Aymara Indians of Bolivia and Peru," J. Ethnolog. Soc. Lond.,

2 (1870): 193-305. Year for letter determined by this publication date.
2. See Darwin, Expression of the Emotions (1872), 232, 318, and 336.

382 187[0][1] Aug. 23.

To [Jean Louis Armand de QUATREFAGES de Breau]; Down (type 7)
LS; 8¼ X 5¼; 3p. (yellow paper)

Hesitates to write because "private interests are as nothing";[1] thanks for interest shown in CD's election to "your Academy [i.e. Academie des Sciences]";[2] thank [Henry] Milne-Edwards for honor conferred upon CD; sees in last number of Revue that Milne-Edwards believes that "existing species are the modified descendants of extinct species. Such an admission seems to me very much more important than whether natural selection has been a more or less efficient means of change; though for my own part I shd never have been able to admit the evolution of species, unless I cd have partly explained to myself how the innumerable & beautiful adaptations, which we see all around us, had originated";[3] while [Jean Baptiste Armand Louis Leonce] Élie de Beaumont calls CD's science "frothy", his own bubbles first of craters of elevation and second of direction of mountain chains according to age have "burst & vanished into thin air" everywhere but France; next book [i.e. Descent of Man (1871)], due in two or three months, will displease corr., but will send copy of it anyway, because corr. knows CD publishes "only what I believe to be the truth, after mature consideration."

1. CD is lamenting the publication of Quatrefage's confidential remarks; see "Darwin before the French Academy," Nature, 2 (1870): 298. Year for letter determined by this publication date, plus those others of 1870 below.
2. See Robert E. Stebbins, "France," in Thomas F. Glick, ed., The Comparative Reception of Darwinism (Austin: Univ. of Texas Press, 1974), passim, esp. 146-48.
3. Milne-Edwards, "Sur les travaux de Ch. Darwin," Revue scient., Paris, 7 (1870): 591.

383 [1870] August 30th

To [Philip Lutley] SCLATER; Down (type 5)
ALS; 6 3/4 X 5; 1p. and end. [136/ sending an/ article on/ a woodpecker]
(enclosure wanting)

Submits enclosed note concerning the woodpecker of the plains to the Society for publication versus [William Henry] Hudson.

384 1870. Octr 6.

To [John Frederick William HERSCHEL][1]; Beckenham/ Kent S.C. [sic]

Copy of L; 7 X $4\frac{1}{2}$; 2p.

Cannot correct the little Manual; is unwell and is leaving for rest; upon return, must correct proofs of work on hand; ask Professor [John] Phillips to revise the essay and append his own name; Phillips's geological knowledge is more up-to-date than CD's.[1]

1. See Darwin, "Geology," in Herschel, ed., A Manual of Scientific Enquiry... (1849), 156-95. In the 1871 edition of this article, the essay was revised by Phillips. When this letter appeared, CD was working on proofs of Descent of Man (1871); see Life and Letters, III, 120.

385 [1870][1] Nov. 15

To [William Darwin] FOX; Down (type 5) (distinctive gray ink used)
ALS; 8 X 5; 4p.

Supposes Fox now at Isle of Wight; winter has come with a vengeance; wife [Emma Wedgwood Darwin] wrote about governesses after receiving letter of October 28; supposes Fox was unable to come to Down; is ill from correcting proofs of two volumes [of Descent of Man (1871)]; owes many facts in book on sexual differences of birds to Fox; will send copy when published, even though Fox will probably disapprove about origin of man; "It is very delightful to me to hear that you, my very old friend, like my other books, & you were one of my earliest masters in Nat[ural] History"; is weak, feels "each job as finished must be my last."

1. Proofs of Descent of Man were corrected in late 1870; see "Darwin's Journal," 18.

386 [1870] Dec. 25th [pmk. DE 26/ 70]

To [Charles] LYELL; Down (type 5)
ALS; 8 X 5; 1p. and env., add. [Sir C. Lyell Bart./ 73 Harley St/ London/ W.], end. [C. Darwin]

Thanks for new book [i.e. Lyell, The Student's Elements of Geology (London: John Murray, 1871), advance copy]; will look at it when finished correcting proofs [of Darwin, Descent of Man (1871)]; has only one short chapter of proofs to go.

387 [1871] March 12th

To [George] BUSK; Down (type 5)
ALS; 8 X 5; 4p.

Thanks for explaining blunder; [George] Rolleston had already written to correct the confounding of the two foramina, so latter 2,000 copies [of Darwin, Descent of Man (1871)] have been corrected, but this is not even noticed in errata of first and larger lot; regarding mistaken notion that supra-condyloid foramen was present in anthropomorphous apes, was led to error because [St. George Jackson] Mivart says it is present in Anthropoidea, and CD incorrectly assumed that Anthropoidea were apes nearest allied to man; will strike out or qualify strongly the passage on supra-condyloid; thinks remarks on what [Thomas Henry] Huxley calls the inter-condyloid are applicable; new reprint or new edition [of Darwin, Descent of Man (1871)] will be out at end of upcoming week; will send a copy to Busk.

388 [1871][1] March 18th

To [Thomas Roscoe Rede STEBBING][1]; Down (type 5)
ALS; 8 X 5; 3p.

Thanks for essays; glad they were published in book;[1] thinks they will do good; esteemed those read some time ago when sent separately; corr. is bold to publish them; corr.'s preface is gratifying; offers corr. a copy of Darwin, Descent of Man [(1871)].

1. See Stebbing, Essays on Darwinism (London: Longmans, Green, and Co., 1871). Year for this letter determined by this publication date. See also letter to Stebbing, March 3, 1869, above. This letter was purchased as a part of a collection with many other letters to Stebbing.

389 1871 March 30th..

From Anne [Henslow] BARNARD; Bartlow Leekhampton Cheltenham
ALS; 7 X 4½; 5p., sketch

Has not seen CD since she was child; Darwin, Descent of Man [(1871)], reminds her of event of interest to CD; on October 20, 1852, visited Colchester idiot asylum with father, Professor [John Stevens] Henslow; saw there a girl with pointed ears, as in accompanying sketch; ears not quite like those of donkey, but similar; girl may still be in asylum.

390 [1871][1] March 31st

To [Anne Henslow] BARNARD; Down (type 5)
ALS; 8 X 5; 1p.

Thanks for letter; has written to doctor of large asylum about pointed ears, but has not yet received reply; "To the last day of my life I shall think of your Father [John Stevens Henslow] with the deepest respect, and affection, & gratitude for his invariable kindness towards me."

1. See preceding letter, above.

391 [1871] March 31st [pmk. MR 31/ 71]

To P. B. MASON; Down (type 5)
ALS; 8 X 5; 4p. and env., add. [P. B. Mason Esqr/ Burton-on-Trent.]

Thanks for prompt information; in next edition of book [?Descent of Man (1871)], will quote statement, new to CD, about hair on back; wants quantitative data on this; Mason's suggestion about size of male infants is good; should be able to find data in Dr. [?James Matthews] Duncan's paper in Trans. R. Soc. Edinb.; Mason's solution cannot apply to "the whole case", because male infants have higher mortality rate than female.

392 [1871][1] April 1.

To [Frederick HARRISON]; Down (type 5)
ALS; 8 X 5; 3p.

Thanks for note and for writing at some length; will consider information when preparing corrected edition [of Darwin, Descent of Man (1871)]; cannot make major revisions now because printers are printing off too quickly; will reiterate caution that "beauty" as used by CD does not mean "the highly complex & lofty notions implied by cultivated [sic; ?cultured] men"; capacity of enjoying beauty is acquired; thinks corr.'s remarks on "quasi-selection of opinions & judgments for admiration by society" are just; "I have made too much of natural selection, though less than some who agree with me generally on such points"; would like to read anything corr. publishes on these subjects.

1. Year determined by date of publication of Descent of Man. Corr. identified by autograph dealer who sold letter.

393 [1871 April 27] [pmk. AP27/ 71]

To [?Jurgen Edward] PFEIFFER; no location
Env. only; address reads "Mr. Pfeiffer/ Mayfield/ West Hill/ Putney/ London S.W."

394 [1871] April 29th [pmk. AP30/ 71]

To [William O.] OGLE; Down (type 5)
ALS; 8 X 5; 3p. and env., add. [Dr. Ogle/ 34 Clarges St./ Piccadilly/ London/ W.]

Printed in full, with minor changes: Life and Letters, III, 143.

395 [?1871][1] May 9th

To ?; Down (type 6)
ALS; 8 X 5; 1p.

Thanks for curious anecdote about dog; will not amplify the part of Darwin, [?Descent of Man (1871)] in question.

1. Year determined by Down address variant used. Of the five years during this variant was used, only in 1871 was the address printed to the right side of the first page of the letterhead, as was done with this letter. In other years, it was printed in the center.

396 [?1871][1] May 10th

To ?; Down (type 6)
ALS; 8 X 5; 2p. (enclosures wanting)

Thanks for extracts, new to CD because he rarely sees a foreign paper; was amused by some extracts; three stanzas about the Ascidians are charming; wants more such extracts if they should appear; encloses two photographs of himself.

1. Year determined by Down address variant used. Address is printed to right, which only happened in 1871.

397 [1871][1] June 1st

To [Edward SUESS]; Down (type 6)
ALS; 8 X 5; 3p.

Acknowledges election as foreign honorary member of "the Imperial Academy" [of Austria]; finds this gratifying after the contempt shown by the Paris Institute [i.e. Academie des Sciences]; sorry corr. is ill from last expedition; thanks for word about "your late celebrated Admiral [Wilhelm von] Tegetthoff."[1]

1. Down address variant used is printed to right side--a combination unique to 1871. Tegetthoff died in 1871. CD was elected a foreign corresponding member of the Imperial Academy of Sciences of Vienna in 1871; see Life and Letters, III, 374. He was not elected an honorary foreign member until 1875.

398 [?1871][1] June 2d

To ?; Down
ALS; 8 X 5; 4p.

Women in labor close the orbicular muscles around their eyelids; does this contraction cause tears to flow towards the end of parturition; does similar contraction during violent retching cause tears; do "'Pulver [Maclers?] Volta-Electric Chain bands'" help in cases of "dyspeptic & nervous weakness"; man who sells this aid quotes Sir C[harles] Locock, H[enry] Holland, [Robert] Ferguson, and others; is it quackery and lies, or is it "worth trying as an experiment?"[1]

1. On the production of tears, see Darwin, Expression of the Emotions (1872), 159ff., esp. 165. This passage was written in 1871, which is reason for year for this letter.

399 [1871] June 22nd [end. June 24 1871; pmk. JU23/ 71]

To J. Traherne MOGGRIDGE; Down (type 6)
LS; 8 X 5; 6p. and env., add. [J. Traherne Moggridge Esq/ 8 Park Hill/ Queen's Road/ Richmond/ Surrey], end. [Ch. Darwin/ recd. at Richmond/ June 24 1871]

Thanks for letter containing new facts; hopes Moggridge continues to observe ants; read [Jean] P[ierre] Huber, Recherches sur les Moeurs des Fourmis [Indigènes] (Paris: 1810); published extract of letter from Texan on ants storing and planting seeds in J. Linn. Soc.; [Richard] Kippist [Linnean's librarian] could find it; glad Moggridge will compile notes on variability of certain plants; thinks fly ophrys is inconspicuous, inodorous, and unattractive to insects because, in most cases, not all its flowers are fertilized; compare seeds of fly ophrys specimens fertilized in natural way by insects to those of plants fertilized thoroughly and artificially to see if latter are larger and more robust; is ill, like Moggridge.

400 [1871] June. 26 [pmk. JU26/ 71]

To M[ichael] FOSTER; 6 Queen Anne St. W.
ALS; 8 X 5; 4p. and env., add. [Dr. M. Foster/ 5. Vanburgh Park Road/ Blackheath.]

Invites corr. to visit CD; returns to Down Friday or Saturday [June 30 or July 1]; will take a month's vacation a fortnight after that;[1] wants some curare, if soluble in water or vegetable acids;[2] wants to know how much to dilute curare; [Thomas Henry] Huxley could not help about curare, so he suggested Foster.

1. CD vacationed at Haredene, Albury Heath; see: Life and Letters, III, 150; Emma Darwin, II, 245; and "Darwin's Journal," 18.
2. See Darwin, Insectivorous Plants (1875), chap. IX passim, esp. 204-05 and 223-24.

401 [1871 June] 29 [pmk. JU29/ 71]

To M/ ichael_7 FOSTER; Down (type 6)
ALS; 8 X 5; 1p. and env., add. / Dr. M. Foster/ 5. Vanburgh Park Road/ Blackheath_7

Thanks for curare; what Foster says about amoebae is "a damper".[1]

1. See preceding letter, above.

402 1871 July 20th

To / John Dean CATON_7; Down (type 6)
ALS; 8 X 5; 2p.

Bearers of this note are CD's two sons / George Howard Darwin and Francis Darwin_7,[1] who are touring / United_7 States; please extend aid to them and show them corr.'s "famous Deer-park";[2] sent copy of Darwin, Descent of Man (1871).

1. See Emma Darwin, II, 250 and 250n.
2. See Caton, The Antelope and Deer of America..., 2nd ed. (New York: Forest and Stream Publishing Company, 1877), esp. ix. Recipient of this letter determined by this reference to deer park.

403 1871 Sep 15.

To J/ ohn_7 Royle MARTIN; Down (type 6)
LS; 8 X 5; 2p. (enclosures wanting)

Thanks for letter of 12th; takes ten shares in Artizan's Dwelling Company; will send check for £ 100.

404 / 1871_7 Oct. 10th

To ?; Down (type 6)
ALS; 8 X 5; 2p.

Experiment suggested by corr. is not worthwhile; any difference in the duration of human bones would have been detected already in old bones found in caverns and tumuli and in mummies; / Thomas Henry_7 Huxley has sent an admirable review of / St. George Jackson_7 Mivart's book to the Contemporary for the November number;[1] is preparing new / sixth_7 edition of Darwin, Origin / (1872)_7.

1. See: Mivart, On the Genesis of Species (London: Macmillan, 1871); and Huxley, "Mr. Darwin's Critics," Contemporary Review, 18 (1871): 443-76, issued November, 1871.

405 / ca. 1871-1875_7[1] Jan. 8th

To ?; Down (type 6)
ALS; 8 X 5; 1p.

Thanks for letter of January 6; declines to join movement mentioned by corr., since CD is ignorant of the arguments on either side of the issue.

1. Years determined by Down address variant used. This method first suggested for this letter by Sydney Smith, St. Catharine's College, Cambridge University.

406 [ca. 1871-1875][1] March 23d

To ?; Down (type 6)
ALS; 8 X 5; 1p.

Lists gold and silver pheasant in table of authentic crosses among the Phasianidae, drawn up several years ago, but has by accident omitted this case [in recent publication]; would not have introduced it without fairly good evidence; hopes corr. will publish case.

1. Years determined by Down address variant used.

407 [ca. 1871-1875][1] June 8th

To [Thomas Campbell] EYTON; Down (type 6)
ALS; 8 X 5; 2p. and end. [1647 to 73/ <u>Original Sketch</u>/ 4th (Amsl.?) 1676/ another 1686]

Thanks for letter; story must be "a cock & a Bull," since [Robert] Swinhoe, who knows Formosa well, would have heard of it otherwise; would like extract to send to Swinhoe, now in China; discusses health.

1. Years determined by Down address variant used.

408 [ca. 1871-1875][1] June 9th

To ?; Down (type 6)
ALS; 8 X 5; 1p.

Would be happy to read paper, but is a poor critic; believes all he reads until long reflection causes doubts; finds such reflection hard work nowadays; has no mathematical knowledge.

1. Years determined by Down address variant used.

409 [ca. 1871-1875][1] July 9th

To [Léo Abram ERRERA]; Down (type 6)
ALS; 8 X 5; 1p.

Thanks for exertions in CD's behalf; regrets waste of corr.'s time.

1. Years determined by Down address variant used.

410 [ca. 1871-1875][1] Nov. 8th

To ?; Down (type 6)
ALS; 8 X 5; 1p.

Orders a "Vulcanized Indian-Rubber Enema of medium size."

1. Years determined by Down address variant used. The following persons helped to date this letter: Sydney Smith, St. Catharine's College, Cambridge University; Sir Hedley Atkins and Philip Titheradge, Down House; and Dr. Ralph Colp, Jr., New York City.

411 [ca. 1872][1] Feb. 4th

To [John Edward] GRAY; Down (type 6)
ALS; 8 X 5; 1p.

Remembers "reading in Molina about the Huemul";[1] asked about it in southern parts of South America, concluded it did not exist; must be rare; glad Gray has made out what it is;[1] regards to Mrs. [Maria Emma] Gray.

1. See: Giovanni Ignazio [i.e. Juan Ignatius] Molina, <u>The Geographical, Natural and Civil History of Chili</u>, 2v., tr. from the original Italian (London: 1809); and Gray, "On the Guémul (<u>Huamela leucotis</u>)," <u>Ann. Mag. nat. Hist.</u>, <u>10</u> (1872): 445-46; and <u>11</u> (1873): 214-20 and 308-10. Discussion is about the guemal, or South American deer. Year for letter determined by date of publication of Gray's paper.

412 1872 March 16.

To Messrs. Appleton & Co.; Down (type 6)
ALS; 7 X 4½; 3p.

Thanks for letter of February 23; is gratified about sale of CD's works; thanks for payment of £ 427·13·8, to be sent by Mr. Layton; is glad Darwin, <u>Journal [of Researches]</u> sells well.

413 1872 Ap 8.

To ?; Down (type 6)
LS; 8 X 5; 2p.

Thanks for paper and kind remarks; had read [Harriet] Martineau's book [The Positive Philosophy of Auguste Comte, freely translated and condensed by H. Martineau (1853)] and others on [Isidore Auguste Marie François Xavier] Comte, but had forgotten Comte's observations on relations of man to the lower forms of life.

414 [1872] April 12th [end. 1872./ April 13.]

To ?; Down (type 6)
ALS; 8 X 5; 1p. and end. [1872./ C. Darwin/ Kent./ April 13 (and) 1872/ C. Darwin/ Down/ Apr 15] and attached enclosure[1]

Orders the following: J[ohn R.] Leifchild, The Higher Ministry of Nature... ([London]: Holder & Stoughton [sic; Staughton], [1872]); Erm Müller, The Application of the Darwinian Theory to Flowers, and the Insects Which Visit Them (Salem, Mass.: Naturalists' Agency, n.d.);[1] and James [Bradbridge] Hunter, "A Review of Darwin's Theory" (New York: Appleton and Co., [1871]), previously ordered, and which CD is anxious to get.[2]

1. Enclosure is a broadside advertisement for Müller's book. Müller's publication is reprinted from Am. Nat., 5 (1871): 271-97. See also Vorzimmer, Reprint Catalogue, item R.212.
2. The review appeared in the Journal of Psychological Medicine; see Vorzimmer, Reprint Catalogue, item R.213.

415 [1872] May 10th

To [Charles] LYELL; Down (type 6)
ALS; 8 X 5; 4p.

Printed in full, with minor changes: More Letters, I, 336-37 (letter 250). p. 336, line 9, and p. 337, line 1, change "a higher grade" to "higher grades".

416 [1872][1] May 22

To [Charles] LYELL; Down (type 6)
ALS; 8 X 5; 2p.

Thanks for letter; see Darwin, Origin [(1872)], new edition, pp. 332 and 348, for remarks "on forms not changing when migrating in a body"; some naturalists think this view true; read Prof. [Friedrich Leopold August] Weismann, Ueber den Einfluss der Isolirung [auf die Artbildung] (Leipzig: [Wilhelm Engelmann], 1872); Lyell has had a grand tour.[1]

1. The Lyells had gone to France; see Life of Lyell, II, 438. This, plus date of publication of sixth edition of Origin, determine year for this letter.

417 1872 May 23

To "Madam"; Down (type 6)
LS; 8 X 5; 4p.

Thanks for letter; will help corr. if possible; wrote to judge at Crystal Palace for dates of reports and information about books on cats; does not know of works exclusively on larger felines; read [Alfred Edmund] Brehm's [Illustrirtes] Thierleben [6v. (Hildburghausen: Bibliographisches Institute, 1864-1869)], which has been translated into French; thanks for information about fertility of crossed Angora and common cats; does not think characters of two fathers can be transmitted to single individual, but new paper by Fritz Müller indicates this is possible; thanks for book, which CD hopes will discuss expression of emotions in cats.

418 1872 June 1

To [Charles] LYELL; Down (type 6)
LS; 8 X 5; 2p. (enclosure wanting)

Thanks for Wood's interesting letter; there is much evidence that varieties of fruit trees transmit characters to a large extent, but Wood's cases exhibit exceptionally close identity; [Joseph] Decaisne has recorded cases of surprising diversity; sterility of Wood's seedlings is most remarkable point; "...if they had been planted separately in good ground, they wd not have been so sterile after early youth"; consider, for example, great fertility of seedling peach trees in Americas and in Australia; returns Wood's letter.

419 [?1872][1] June 6th

To [?Michael FOSTER]; Down (type 6)
LS; 8 X 5; 3p.

Has read corr.'s little essay; it tells exactly what CD wanted to know, namely, how far our knowledge extended and where we were ignorant; invites corr. to Down; some of CD's sons will be home in week or two; is ill, so can only talk for short periods.

1. Year determined by Sydney Smith, St. Catharine's College, Cambridge University; address, however, is printed to right side of page, which with this Down address variant usually indicates a year of 1871.

420 [1872] July 12th

To [Charles] LYELL; Down (type 6)
ALS; 8 X 5; 4p.

Printed, with minor changes and minor omissions: More Letters, II, 164-65 (letter 511).

421 1872 July 27th Saturday

From John T[homas] GULICK; c/o Mrs Delacour/ Frindsbury Hill/ Rochester/ Kent
ALS; 7 X $4\frac{1}{2}$; 4p.

CD's account of Galapagos Islands [in Darwin, Journal of Researches (1839)], read long ago, led Gulick to study distribution of species at Sandwich Islands; see Gulick's article in July 18 number of Nature; is visiting England; wants to meet CD and show him some land shells from Sandwich Islands, "illustrating the extremely limited distribution of the species, and the continuous gradation of forms from one species into another"; will leave soon for North China and Mongolia.

422 [1872] July 28th [pmk. JY29/ 72]

To J[ohn] T[homas] GULICK: Down (type 6)
ALS; 8 X 5; 4p. and env., add. [J. T. Gulick Esqr--/ care of Mrs. Delacour/ Frindsbury Hill/ Rochester], end. [From Darwin]

Printed in full, with minor changes: Addison Gulick, Evolutionist and Missionary: John Thomas Gulick... (Chicago: Univ. of Chicago Press, 1932), 233-34. p. 234, line 1, missing name is "Prof. [Franciscus Cornelius] Donders of Utrecht (whom I could not refuse to see)".

423 [1872] Aug. 8th [pmk. AU8/ 72]

To J[ohn] T[homas] GULICK: Down (type 6)
ALS; 8 X 5; 3p. and env., add. [The Revd. J. T. Gulick/ at Mrs. Delecour/ Frindsbury Hill/ Rochester.], end. [From Darwin]

Printed, with minor changes: Addison Gulick, Evolutionist and Missionary: John Thomas Gulick (Chicago: Univ. of Chicago Press, 1932), 234-35. At beginning of letter is: notes crossed on the road; cannot do anything about Gulick's paper;[1] the secretaries act like judges and demur to any suggestions being made to them; is prejudiced in favor of Gulick's subject. p. 235, line 1, change "Section" to "[Secys.?]".

1. Refers, perhaps, to Gulick, "Diversity of Evolution under One Set of External Conditions," Rep. Br. Ass. Advmt Sci., 42 (1872), pt. 2: 136. Full paper printed in J. Linn. Soc. (Zool.), 11 (1873): 496-505.

424 [1872] Aug. 30th [pmk. AU30/ 72]

To [Joseph Dalton] HOOKER; Down
ANS on p.c.; 3 X 4 3/4; 1p. and add. [Dr. Hooker F.R.S./ Kew/ London W.]

Acknowledges check and note; will not send it to bank until after talking with Hooker.

425 1872 Nov 21.

To [John Maurice] HERBERT; Down (type 6)
LS; 8 X 5; 4p.

Has just published Darwin, Expression of the Emotions (1872); has sent Herbert a copy; recalls "jolly old days" at Barmouth and Cambridge;[1] does Herbert remember anonymously giving CD a microscope; no other event in CD's life "surprised & gratified me more"; is "a confirmed invalid"; wants news of [Charles Thomas] Whitley; has seen S[amuel] Butler, author of Erewhon... [(London: 1872)] and son of Tom [i.e. Thomas] Butler; latter has become "a very unpleasant old man."

1. See Life and Letters, I, 165-66. Concerning the microscope, see Gavin deBeer, ed., "The Darwin Letters at Shrewsbury School," Notes Rec. R. Soc. Lond., 23 (1968): 73.

426 1873 Ap. 7

To ?; Down (type 6)
LS; 8 X 5; 2p.

Thanks for invitation and for catalogue of corr.'s collection of Cretacean fossils; would like to see Brighton Aquarium, but poor health forbids it; accepts corr.'s curious instance of inheritance.

427 [1873][1] May 16th

To [Charles] LYELL; Down (type 6)
ALS; 8 X 5; 2p.

Thanks for copy of fourth edition of Lyell, [The Geological Evidences of the] Antiquity of Man [(London: John Murray, 1873)]; will read over the modified or new parts; book seems larger than earlier editions.

1. Year determined from date of publication of Lyell's book.

428 [1873] June 7th [end. June 10/ 73; pmk. JU7/ 73]

To G[eorge] CUPPLES; Down (type 6)
ALS; 8 X 5; 4p. and env., add. [G. Cupples Esqr/ The Cottage/ Guard Bridge/ Fifeshire/ N.B.], end. [Mr Darwin/ Recd. June 10/ 73]

Thanks for letter, for former letter, and for report of [Julius] Victor Carus's lecture; is away from home on rest trip;[1] glad CD's suspicion about [James Hutchinson] Stirling was groundless; was mortified that "so able a man" wrote "with such loathing contempt of me"; has not seen [Ralph Waldo] Emerson; hears that Emerson is charming, but probably does not have much in common with him; "During the last 2 or 3 years we have seen several Yankees, & as a rule they seem a most pleasant set"; was charmed with the Nortons [Charles Eliot Norton and his wife Susan Ridley Sedgwick Norton]; has recommended "Tappy's chickens" to CD's hosts.[2]

1. CD went to Leith Hill Place from June 4 to 12; see "Darwin's Journal," 19.
2. Anne J. Cupples, Tappy's Chicks and Other Links between Nature and Human Nature... (London: Strahan & Co., 1872). Anne J. Cupples was the wife of corr.

429 1873 June 23.

To Dr. [Edward Wickstead] LANE; Down (type 6)
L (postscript in CD's hand); 8 X 5; 3p. (mutilated)

Thanks for copy of Lane's little book [Old Medicine and New (London: 1873)]; will read it soon; owes much to hydropathy, even though it seemed to do harm the last time it was tried; had a pleasant time whenever visiting Moor Park; regards to wife and to "Lady Drysdale [Lane's mother-in-law]".[1]

1. See Emma Darwin, II, 184.

430 [1873][1] July 18th

To ?; Down (type 6)
ALS; 8 X 5; 1p.

Is not anatomist enought to say whether corr.'s power is very unusual; moving of the scalp voluntarily is not very important, unless very extreme and inherited.

1. Year written in pencil upon original in unknown hand.

431 [1873][1] Sept 5th

To [a chemist, but not Hopkin & Williams]; Down (type 6)
ALS; $8\frac{1}{2}$ X 5; 2p.

Thinks chlorides, instead of nitrates, of the various metals would be better for CD's purpose, but only when such are soluble; "But it is perhaps too late & nitrates wd. do very well, & are necessary in the case of silver"; send bottle of "Oxley's Essence of Ginger"; send four acids when ready, since CD wants to try them before the metallic salts; send the latter in two lots if it takes long to prepare them.

1. Date and recipient suggested by Sydney Smith, St. Catharine's College, Cambridge University. This letter follows one dated the previous day, the original of which is now in possession of Baird & Tatlock (London) Ltd., incorporating Hopkin & Williams; copy on file at APS, courtesy of Baird & Tatlock.

432 1873 Sep. 24.

To [Charles] LYELL; Down (type 6)
LS; 8 X 5; 4p.

Thanks for note and for apples, received "only today, as they were directed to Beckenham instead of to Orpington Station"; seedlings from same parent are "wonderfully alike in fruit and leaves; forgets what [Joseph Dalton] Hooker says; reversions to crab-state are exceptional;[1] wild crabs themselves differ much; see Darwin, Variation under Domestication [(1868)], I, 350, and II, 31; inheritance is not as general with apple trees as Wood thinks;[2] no one has raised a new Ribstone or Golden Pippen,[3] although seedling by [Thomas] Andrew Knight approaches latter; Wood's most remarkable statement is about sterility of seedlings, since they are several years old;[4] it is remarkable that parent trees were not intercrossed, since many flowered simultaneously; if bees carry pollen between trees, then pollen of above-named varieties is prepotent over others, so they have the character of species; would like to fertilize flowers of Hawthornden with pollen from many distinct varieties and to compare resulting seedlings; is ill, under care of [Andrew] Clark; invites Lyell to Down.

1. Lyell has written in pencil in margin here: "This is my contention".
2. See letter to Lyell dated June 1, 1872, above.
3. Lyell has written here: "but how has the failure occurred. probably from sterility or want of vitality".
4. Lyell has written here: "(14th year from the first sown)".

433 1873. Dec 19.

To [A. MOSCHKAU]; Down (type 6)
LS; 8 X 5; 2p. (enclosure wanting)

Thanks for letter; returns the enclosed; has alluded to parrots speaking in Darwin, Descent of Man [(1871)];[1] in new edition, has added good evidence that they understand the words used; will add, on corr.'s authority, the case of the starling; is ill and overwhelmed with letters, so cannot correspond with anyone.

1. See Darwin, Descent of Man (1871), I, 236; and II, 335.

434 [?1873-1875][1] Oct 19th

To A. [G.] DEW-SMITH; Down (type 6)
ALS; 8 X 5; 4p. and env., add. [A. Dew Smith Esqr--/ 24. Green St/ Cambridge]

Son is going to Cambridge tomorrow; takes opportunity to send specimen of Dionaea so corr. can observe changes in temperature in plant when it is in action; gives instructions to assure proper measurement; wishes plant were a better specimen, "but I have given away all by best specimens for [John Scott] B[urdon-]Sanderson's electrical experiments, which have proved so wonderful."[2]

1. Down address variant used determines limits of 1871 and 1875. Fact that address is printed in center of page upon paper having a distinctive escutcheon watermark eliminates 1871 and 1872.
2. See Darwin, Insectivorous Plants (1875), 318.

435 1874 Jan 8th

To [James Philip Mansel WEALE]; Down (type 7)
LS; 9 X $8\frac{1}{4}$; 1p.

What proportion of leaves of Pinguicola [sic; Pinguicula] have insects adhering to them; send leaves having captured insects, so CD can identify insects caught; observe how secretions from captured insects flow around leaf; do leaves ever capture seeds; if so, send specimens; describe habitat, esp. nearby vegetation, for Pinguicula vulgaris.

436 [1874][1] Jan 10

To Dr. [Andrew] CLARK; 6 Queen Anne St
LS; 7 X $4\frac{1}{2}$; 2p.

Has just come to London; please call and see CD.

1. Date determined by Sydney Smith, St. Catharine's College, Cambridge University.

437 [1874 ca. February 20] [end. Recd on 20th Febry 1874]

To ?O.P.L.; Down (type 7)
ALS; 9 X $7\frac{1}{4}$; 4p., end. [Recd. on 20th Febry 1874. O.P.L.]

Thanks for reference, which CD had lost and is delighted to recover and to quote; will discuss point mooted in corr.'s letter; all organs necessary for reproduction are attributed to natural selection; sexual selection occurs only when advantage is gained by struggle between two or more individuals of same sex; males vary more than females; corr.

may consider this a result of masculine "'vital force'", but CD prefers to speak of it as mere variability; this variability does not account for well-marked sexual differences unless variations are accumulated; when sexes are alike, there is no evidence of sexual selection, unless two sexes closely resemble one sex of closely allied form in which sexes differ; if sexes differ much in color, esp. if male is more brilliant, then brightness of male is probably due to sexual selection; evidence of action of sexual selection is only good if one sex displays its colors to the other; individual would not display colors unless this were an advantage, and if it is advantage, this is basis for sexual selection; corr. probably does not think stridulatory organs of insects result from vital force; if such organs result from sexual selection, why not color; [Giovanni] Canestrini says in recent Italian paper that female spiders select a mate from among many males; collected spiders during voyage of Beagle and deposited them in British Museum.

438 [1874][1] March 11

To [Alfred William BENNETT][1]; Down (type 6)
ALS; 8 X 5; 1p.

Has just heard that corr. has ceased to assist editor of Nature; where are wood blocks of climbing plants, which belong to Linnean Society;[2] must borrow them before long, and they might get mislaid at Macmillan's.

1. For year and recipient of this letter, see: A. J. Meadows, Science and Controversy: A Biography of Sir Norman Lockyer (Cambridge, Mass.: The M.I.T. Press, 1972), 31-32; and DNB, second supplement, 1, 143-44.
2. CD refers to the thirteen illustrations for Darwin, Climbing Plants (1865).

439 [1874][1] Mar 31 [wmk. 1868]

To [John] PHILLIPS; Down (type 7) (black border)[1]
LS; 7 X 4½; 2p.

Thanks for invitation, but journey to Oxford would "nearly annihilate" CD; corr.'s sketches indicate only a big theory would equal corr.'s fossils;[1] will watch for published account; remembers pleasant hours spent with corr. at York.[1]

1. This letter is, apparently, a reply to a letter to CD dated March 14, 1874, which is located in the Robin Darwin Deposit, University Library, Cambridge; the earlier letter contains sketches. The black border may result from death of CD's wife's sister-in-law, Frances Mosley Wedgwood; see Emma Darwin, I, xxvii. The trip to York was in 1845; see Life and Letters, I, 343-44. Information provided by Sydney Smith, St. Catharine's College, Cambridge University.

440 [1874] Ap. 13th [end. 1874/ April 16]

To ?; Down (type 6)
ALS; 8 X 5; 1p. and end. [1874/ Ch. Darwin Esq/ Down/ April 16] (enclosure wanting)

Encloses check; wants the "Flora" (German), for the year 1873, numbers 28 and 29; will accept entire volume if necessary.

441 [1874] Ap. 18th [pmk. AP18/ 74]

To Leonard RUDD; Down (type 6)
ALS; 8 X 5; 1p. and fragment of env., add. [Leonard Rudd Esq/ Guy's Hospital/ London/ S.E.]

Thanks for case of additional mammae; abnormality of this sort is not rare; that mammae enlarge every month is quite new, but does not understand its meaning; this does not occur commonly with women.

442 [1874][1] May 31st

To [Charles] LYELL; Down (type 6)
ALS; 8 X 5; 1p.

Thanks for note and that of [George Julius Poulett] Scrope; "What you have done[1] may be the first step in an admirable plan"; thinks bequest shows wisdom and would do same, if CD had fewer sons; enjoyed Lyell's visit to Down; is glad Lyell sent [Anton Felix] Dohrn's paper to Scrope.

1. Lyell was bequeathing a large sum to science; see Life of Lyell, II, 477-78. Year determined from this information.

443 [1874] July 13th

To ?; Down (type 6)
ALS; 8 X 5; 2p.

Utricularia arrived safely; was unwell, so son [?Francis Darwin] took charge of specimens and worked at bladders; it will be difficult to make out the function of parts; will his great experience on Desmids under the microscope, corr. knows that one's brain must be clear and hand steady to do anything on difficult structures.

444 1874 July 16

To [George John ROMANES]; Down (type 6)
LS; 8 X 5; 4p.

Printed in full, with changes: More Letters, I, 354 (letter 264). line 1, change "your kind and long communication" to "the copy of your long letter".[1] line 4, change "much" to "at present". lines 7/8, change "some later time, when I may again take up the subject. Your letter" to "a future time. It". line 11, change "through so-called" to "through mere so-called". line 16, illegible word is "severe".

1. For long letter from Romanes, see More Letters, I, 352-54 (letter 263).

445 [1874 July][1] 27th

To [Bartholomew] PRICE; Abinger Hall/ Wotton/ Surrey[1]
ALS; 8 X 5; 2p.

Thanks for specimen of Utricularia, which CD presumes is from corr.; obtained microscope at Abinger, concludes that there is essential similarity [between specimen received and some other specimens], but with some important differences; five of the largest bladders contained dead Entomostraca: 24, 20, 15, 10, and 7 victims respectively; "What slaughter!"; leaves on Thursday [July 30] for three weeks at house of son [William Erasmus Darwin] at Bassett, Southampton.

1. Down address variant, type 6, has been crossed out. Date for letter determined by Abinger address; see "Darwin's Journal," 19-21.

446 [1874][1] July 28th

To [George John ROMANES]; Abinger Hall/ Wotton. Surrey (Post Town)/ Gomshall (Station) S.E.R. [embossed][1]
ALS; 7 X 4½; 1p.

Thanks for letters; "I have so poor a metaphysical head that Mr [Herbert] Spencer's terms of equilibration &c always bother me & make everything less clear".

1. Year determined by Abinger address; see "Darwin's Journal," 19-21.

447 [1874][1] Aug. 7th

To ?; Bassett/ Southampton.[1]
ALS; 8 X 5; 1p. (mutilated)

Is glad hop interests corr.; numerous almost small angular bodies, which CD imagined to be ovules, are worth corr.'s attention; would be proud to appear in an engraving in Gardeners' Chron.; is taking a three-week rest.

1. Year determined by Bassett address. Only year in which CD was there on August 7 was 1874; see "Darwin's Journal," 19. Down address variant, type 6, is crossed out.

448 [1874] Sept. 3d

To [Charles] LYELL; Down (type 6)
ALS; 8 X 5; 4p.

Printed in full, with minor changes: More Letters, II, 237-38 (letter 571).

449 1874 Sep 7.

To Lady Dorothy [Fanny Walpole] NEVILL; Down (type 6)
LS; 8 X 5; 3p.

Thanks for Drosera plant; does corr. have specimen of epiphytic Utricularia, which produces minute bladders only when making fresh shoots or leaves?

450 1874 Sep 23.

To [Charles] LYELL; Down (type 6)
LS (postscript in CD's hand); 8 X 5; 3p.

Printed, with minor changes: Life and Letters, III, 190. line 14, missing sentence is: "I have just begun Mr [Thomas Francis] Jamieson's paper,[1] which seems to me very interesting & I shd think true." At end of letter is: [Robert] Mallet will not like to hear of palaeozoic true sub-aerial explosive volcanoes.

1. "On the Last Stage of the Glacial Period in North Britain," Q. Jl. geol. Soc. Lond., 30 (1874): 317-38.

451 1874 Oct. 8th

To [?James Crichton BROWNE][1]; Down (type 6)
ALS; 8 X 5; 1p.

Thanks for West Riding Asylum Report; notes excellent article and photograph by corr.[1]

1. Browne was director of the West Riding Pauper Lunatic Asylum in Yorkshire and editor of the Medical Reports of the West Riding Lunatic Asylum. CD owned at least volumes 1, 2, and 5; see Darwin Library: List of Books Received in the University Library, Cambridge, March-May 1961 (Cambridge: University Library, Cambridge, 1961), under "West Riding".

452 1874 Oct 14th

To [Daniel] OLIVER; Down (type 6)
LS; 8 X 5; 2p.

Thanks for aid and information; plants arrived safe; will set to work in two days and will then return them; thank [Joseph Dalton] Hooker when he returns.

453 [?1874 December 7][1] Monday night

To [George John ROMANES]; 2 Bryanston St/ Portman St[1]
ALS; 8 X 5; 1p.

Would like to meet corr.; planned to visit Brown Institution; come for lunch on Wednesday.

1. Down address variant, type 8, is crossed out. According to "Darwin's Journal," 19-21, there were eleven weeks in which CD was at Bryanston Street address on both Monday and Wednesday, between 1873 (first possible year of correspondence with Romanes) and 1882 (year of CD's death). Type 8 address variant was not employed until at least late 1874. CD begins greeting Romanes by name in salutations to letters on December 16, 1874, but this letter begins with "My dear sir". Letter is thus dated with the only candidate Monday in late 1874, namely December 7. This may indicate first use of type 8 variant.

454 [?1874 December 9][1] Wednesday evening

To [George John ROMANES]; 2. Bryanston St[1]
ALS; 8 X 5; 1p.

Wants corr. to call tomorrow morning, hopes to be well then.

1. Down address variant, type 8, is crossed out. This letter appears to follow after the preceding letter, above. In this letter, CD still greets Romanes formally in the salutation.

455 1874 Dec 16th

To [George John] ROMANES; Down.
LS (postscript in CD's hand); 9 X 7¼; 4p.

Thanks for Romanes's book [?Christian Prayer and General Laws... (London: 1874)]; already has copy from Macmillan; diagram is excellent, as is reasoning; must think over subject; Romanes must show advantage of longevity; diagram may apply to sterility of distinct species, although CD has argued elsewhere [Variation under Domestication (1868), II, 185-89] that intersterility is incidentally acquired; glad Romanes is taking up pangenesis; vine with digitate leaves is good for graft-hybrids; asked [Thomas Henry] Farrer's gardener to raise such a plant; read Variation under Domestication, I, 395, on vines; facts are from [Carl Friedrich von] Gärtner's [Versuche und Beobachtungen über die] Bastarderzeugung [im Pflanzenreich (Stuttgart: K. F. Hering, 1849), 619f.]; does not have [?H. Adorne de] [Tscharner's?] or [Georg Heinrich] Ritter's book;

Horticultural Society, Royal Society, and Linnean Society have good libraries in scientific horticulture; plants will serve better than animals; suggests hyacinths, dahlias, crocuses, potato, and beet for graft-hybrid experiments, but disclaims expertise; would consult [Joseph Dalton] Hooker, but wife [Frances Harriet Henslow Hooker] just died; will find consultant at Kew Gardens; Romanes should prepare for "endless disappointments, as your first experiments will merely serve to teach you what ultimately must be done"; may have seen account of successful grafts of two colored beets; "It wd. be very difficult to get people to see vast importance of graft-hybrids, as throwing light on sexual generation."

456 1874 Dec 23.

To [George John] ROMANES; Down (type 8)
LS; 8 X 5; 1p. (enclosure wanting)

As enclosed shows, [Joseph Dalton] Hooker wishes to help Romanes, so call at Kew soon.[1]

1. See preceding letter, above. Romanes had visited Hooker by January 14, 1875; see Life of Romanes, 19-20.

457 1874 Dec 27th

To [George John] ROMANES; Down (type 8)
LS; 8 X 5; 2p.

Is glad the letter from [Joseph Dalton] Hooker pleases Romanes;[1] Hooker is "the best & kindest man I have ever known"; grafting of ears of rabbits would be almost impossible, since rabbits would not remain quiet afterwards; supposes Romanes will use chloroform; comb of fowl, especially Spanish breed, is strongly inherited and thus well fitted for experiment; supposes birds could be chloroformed.

1. See preceding letter, above.

458 [ca. 1874][1]

To [William Bernhard] TEGETMEIER; no location
Printed form letter, with autograph insertions, S; 8 X 5; 3p., add. [Mr. Tegetmeier,/ 346, Strand,/ London, W.C.]

Orders Boddaert's Table des Planches Enluminees.

1. Date determined by Sydney Smith, St. Catharine's College, Cambridge University. A letter from CD to Tegetmeier dated April 5, 1874, on deposit at New York Botanical Gardens, begins with "Many thanks for Boddart."

459 [ca. 1874-1879][1] Nov. 4th

To [?William] MARSHALL; Down (type 9)
ALS; 8 X 5; 1p.

Miserable summer caused acorns to drop off cork tree; will consult J[oseph Dalton] Hooker to see whether *Quercus rubra* or *Q. coccinea* is handsomer and will have one sent to corr. "for my own honour & glory."

1. Down address variant used determines absolute endpoints of 1874 and 1882. CD died in April, 1882. CD's handwriting became decidedly shakier in 1880, and this letter is in a steady hand.

460 [ca. 1874-1880][1] Friday

To [William O.] OGLE; Bryanston St--Portman St[1]
ALS; 8 X 5; 1p.

Called on Ogle in morning, but he was out; invites corr. to lunch on Sunday.

1. Down address variant, type 8, is crossed out. This letter is written in steady hand; shaky hand began sometime in 1880.

461 [1875][1] Jan 13th

To [George John] ROMANES; Down (type 8)
LS; 8 X 5; 1p.

Has received German journal from Berlin containing several accounts of graft-hybrids produced not by splicing but by inserting bud of one potato tuber into another kind of tuber; results are clear and are compared with sexual generation; offers journal from Germany to Romanes.

1. For Romanes's reply, see *Life of Romanes*, 19-20.

462 1875 Jan 19th

To [A. G. DEW-SMITH]; Down (type 8)
LS; 8 X 5; 2p.

Thanks for letter with account of Zoological Station in Naples; tell [Michael] Foster to publish [Anton] Dohrn's letter and circular as they stand; publish list of donors and amounts subscribed; remembers disappointment in past when he subscribed to something and then did not hear any news of the project's success.

463 1875 Feb. 15

To Lady Dorothy /‾Fanny Walpole_7 NEVILL; Down (type 8)
LS; 8 X 5; 2p.

Thanks for note; will not be in London until late spring /‾actually, March 31_7; is busy on Insectivorous Plants /‾(1875)_7; just acknowledged corr.'s aid in chapter on Utricularia / page 431_7, proofed in morning.

464 1875 March 7

To ?; Down (type 8)
LS; 8 X 5; 1p.

Thinks corr. wrote article on CD in Gardeners' Chron.; thanks for the honor.

465 /‾1875_7[1] Ap. 7th

To /‾George John_7 ROMANES; 2 Bryanston St.[1]
ALS; 8 X 5; 3p.

/‾Thomas Henry_7 Farrer's gardener has cut-leaved vines almost ready;[2] will send them to Romanes in Scotland; /‾Thomas Henry_7 Huxley and Lawson Tait will be at Down on 18th; invites Romanes to come to meet them on Saturday the 17th; do not discuss experiments on animals before Darwin women, since it would horrify them.

1. Down variant address, type 8, is crossed out on top, and a little hand is drawn in which points to the lower half of the printed address. The only year in which CD was at Bryanston Street on April 7 was 1875; see "Darwin's Journal," esp. 19. In addition, April 17, 1875, was a Saturday.
2. See letter to Romanes, December 16, 1874, above.

466 /‾1875 May 2_7 /‾pmk. MY2/ 75_7

To J/‾ohn_7 Jenner WEIR; no location
Address leaf only; address reads: "J. Jenner Weir/ Blackheath".

467 1875 May 3

To /‾James_7 PAGET; Down (type 8)
LS; 8 X 5; 1p.

Thanks for lectures /‾Paget, Clinical Lectures and Essays..., ed. by H. Marsh (London: 1875)_7; vivisection question goes on as well as could be desired.

468 1875 May 19th

To ?; Down (type 8)
ALS; 8 X 5; 1p.

Has had letter from Prof. [August] Weismann, who is interested in corr.'s experiments on birds not eating gaudy caterpillars; has corr. published more lince papers in *Entomological Transactions*; does corr. know of any others working on this subject, such as [Charles Valentine] Riley of United States; Riley is probably the best.

469 1875 May 29

To ?; Down (type 8)
LS; 8 X 5; 2p.

Thanks for paper on ostrich feathers; sends ostrich feather just received from President of Transvaal.

470 [1875][1] June 14th

To Fritz SCHULTZE; Down (type 8)
ALS; 8 X 5; 2p. and add. [An Prof. Dr. Fritz Schultze/ Dresden-(Planen?)/ Reisewitzerstrasse St. (*sic*)]

Thanks for letter and book [*Kant und Darwin...* (Jena: 1875)]; is honored to be coupled with "the illustrious [Immanuel] Kant"; reads German poorly; give CD's regards to [Ernst Heinrich Philipp August] Häckel [i.e. Haeckel] and thank him for note.

1. Year determined from publication date of corr.'s book.

471 [1875] July 12th

To [George John] ROMANES; Down (type 8)
ALS; 8 X 5; 4p.

Printed in full, with minor changes: *More Letters*, I, 359-60 (letter 270).
p. 359, line 14, change "gave" to "saw".
Also printed in full, with minor changes: *Life of Romanes*, 38-39.

472 1875 July 18

To [George John] ROMANES; Down (type 8)
LS; 8 X 5; 3p.

Printed in full, with minor changes: *Life of Romanes*, 32.

473 [1875] August 15th

Figure 7. Charles Robert Darwin, late in life. Photograph by Elliott & Fry; reproduced by permission of Bassano & Vandyk Studios, incorporating Elliott & Fry, London. (American Philosophical Society)

To Mrs. DOWIE; Down (type 8)
ALS; 8 X 5; 3p.

Thanks for note, but [James] Paget says all amputated digits, not just supernumerary ones, tend to regrow a little in the young; must correct erroneous statement; owes correction to corr. and to Paget.

474 [1875 September][1] 11th

To [George John] ROMANES; Basset Southampton
ALS; 7 X 4½; 1p.

The Huxleys [Thomas Henry Huxley and his wife, Henrietta Anne Heathorn Huxley] come to Down on 18th for the Sunday; invites Romanes to meet them and the Litchfields [Richard Buckley Litchfield and his wife, Henrietta Emma Darwin Litchfield]; Frank [i.e. Francis Darwin] may or may not have returned by then.

1. Only twice (on May 11, 1878, and on September 11, 1875) was CD at Bassett on the eleventh of the month and then back at Down the following eighteenth, a Saturday. The earlier date is preferred because Romanes had not yet met the Huxleys at time this letter was written. See "Darwin's Journal," 19-21.

475 1875 Sep 17.

To Dr. S. NEWINGTON; Down (type 8) (last two lines crossed out)
LS; 7½ X 4 3/4; 3p.

Thanks for letter of September 2, explaining about vines; thinks Madresfield vine was a sport of the Hamburgh and reverted to it after being in-arched; has not yet received specimen of berries; had not heard of corr.'s observation of roots secreting carbonic acid, discovered by Dr. N. 25 years since and at first denied by [John] Lindley; Prof. Lacho has proved this by growing plants in polished marble pots; returns J[ohn] Herschel's letter; observations on coincidence of pulse and step are new to CD.

476 [1875] Sept. 24th

To [George John] ROMANES; Down (type 9) (next to last line crossed out)
ALS; 8 X 5; 4p.

Printed, with minor changes and minor omissions: Life of Romanes, 34.
At end of letter is: is glad about corr.'s success with medusae.[1]

1. See Romanes, "Preliminary Observations on the Locomotor System of Medusae," Phil. Trans. R. Soc., 166 (1876): 269-313.

477 1875 Oct 8.

To [George John] ROMANES; Down (type 8) (last two lines crossed out)
LS; 8 X 5; 6p.

Despatched certificate [proposing Romanes for election to Linnean Society][1] to Kew, but it has not returned; will keep it for [Thomas Henry] Huxley's signature when latter visits Down in eight days; potatoes arrived yesterday;[1] on basis of accounts by Germans, thinks two lots are hybridized; inspect large numbers of parent tubers for signs of hybrid mottling; give quantitative data on number inspected; Romanes, rather than CD, should plant the hybrids; does not know rate of reversion to pure forms; no account of same has been published; should potatoes be sent to Scotland;[1] is sorry about onions, "as the nature of seminal hybrids would be so absolutely new";[1] thanks for drawing of striped horse, but subject is finished.[2]

1. On election, potatoes, Scotland address, and onions, see Life of Romanes, 13, 15, 34-38, and 40-41.
2. See Darwin, Variation under Domestication (1875), I, 58-63; and II, 343. This edition was sent to printers on October 3, 1875; see "Darwin's Journal," 20.

478 [1875 November 4][1] Thursday 8th [sic]

To [George John] ROMANES; 6 Q. Anne St.
ALS; 7 X 4½; 4p.

Came to London yesterday to be examined by Vivisection Commission, but returns to Down today; hopes to receive carrots soon; plants will not seed during winter, so wait until spring; mere diffusion of color from side to side [in graft-hybrid carrots] is not important; wants to see if seed yields white and red or mottled carrots; wants seeds of pure forms for control group; saw [James Scott] B[urdon-]Sanderson, who is first to comprehend physiological importance of proving graft-hybrids; latter told CD about medusas; corr.'s papers in Nature, esp. the last,[2] are curious and amusing; part about ferrets was funny and important; it is too late about sternums, but has added a note;[3] will notify Romanes upon receipt of carrots.[4]

1. The "8" written by CD is incorrect. CD testified before the Vivisection Commision on November 3, 1875; see Great Britain, Parliament, Sessional Papers, 1876, v. 41, "Report of the Royal Commission on the Practice of Subjecting Live Animals to Experiments for Scientific Purposes, with Minutes of the Evidence Presented to the Commission," [Cd. 1397], 233-34. Jeffrey L. Sturchio of the University of Pennsylvania provided this reference. For further evidence, this is a reply to letter printed in Life of Romanes, 42-45, and a sort of postscript to this letter is abstracted below.
2. "Instinct and Acquisition," Nature, Lond., 12 (1875): 553-54.

3. See: Life of Romanes, 44; and Darwin, Variation under Domestication (1875), I, 288.
4. See subsequent letter, below.

479 1875 Nov 4.

To [George John] ROMANES; Down (type 8)
LS; 8 X 5; 2p.

Sends postscript to letter sent in morning; carrots have arrived; has potted them and put them in greenhouse because they are so small.

480 1875 Nov. 27th

To Secretary Royal Socy [George Gabriel STOKES]; Down (type 8)
ALS; 8 X 5; 1p. (enclosure wanting)

Submits enclosed paper by [Robert] Lawson Tait.

481 [1875 December][1] 17th

To [George John] ROMANES; 2. Bryanston St[1]
ALS; 8 X 5; 2p.

If skins of birds can be transplanted easily, experiment on spots and barbs by transplanting bit of skin of blue-feathered pigeon; pigeons are easily kept; has been successful in canvassing for [Edwin] Ray Lankester and has excited indignation about his case; returns to Down on Monday; hopes corr.'s paper went off well last night;[2] thinks it a grand discovery.

1. Only time CD was at Bryanston Street on a 17th and returned to Down on Monday was during December of 1875. Down address variant, type 8, is crossed out.
2. Romanes, "Preliminary Observations on the Locomotor System of Medusae [the Croonian Lecture, read December 16, 1875]," Phil. Trans. R. Soc., 166 (1876): 269-313.

482 1875 Dec 26.

To [George John] ROMANES; Down (type 8)
LS (postscript in CD's hand); 8 X 5; 3p.

Printed in full, with minor changes: Life of Romanes, 47-48.

483 1875 Dec 26

To ?; Down (type 8) (last two lines crossed out)
LS; 8 X 5; 1p. (enclosure wanting)

Sorry for "so soon troubling you again"; encloses list of errata for Darwin, Insectivorous Plants [(1875)] detected by [Julius] Victor Carus in course of translating; forward them to French translator.

484 [ca. 1875-1879][1] Sunday Evening

To [George John] ROMANES; 6 Q. Anne St.[1]
ALS; 8 X 5; 1p.

Will call on Romanes early on Tuesday morning; goes to Kew tomorrow.

1. Down address variant, type 8, is crossed out. During the period during which this letterhead was used, there were only six occasions when CD was at Queen Anne Street on both Sunday and Tuesday; the earliest of these is April 4, 1875, and the latest is December 7, 1879. See "Darwin's Journal," 19-21.

485 1876 Fob. 15

To [William O. OGLE]; Down (type 8)
LS; 8 X 5; 1p.

Thanks for book not yet received; send it to 6 Queen Anne St., Cavendish Square; will read it sometime in future.

486 [1876 March 1?][1] [pmk. MAR/24]

To H[enr]y EDWARDS, Jr.; Down (type 8)
ALS; 8 X 5; 2p. and fragment of env., add. [Hy. Edwards Esqr/ San. Francisco/ California/ U. States]

Had received photograph and paper, but was overwhelmed with letters; read corr.'s discussion of Prof. [August Friedrich Leopold] Weissmann's views on dimorphism in butterflies; glad corr. approved of Weissmann's essay, which struck CD much.

1. This date is written on original in ink, in an unknown hand.

487 1876 Ap. 14.

To [George John] ROMANES; Down (type 8)
LS; 8 X 5; 1p.

Will probably receive the paper tomorrow;[1] goes to London on 20th;[2] will call upon corr. shortly thereafter.

1. Probably "An Account of Some New Species, Varieties, and Monstrous Forms of Medusae," J. Linn. Soc. (Zoology), 12 (1876): 524-31.
2. CD was in London April 27 to May 3, 1876; see "Darwin's Journal," 20.

488 [1876] Apl. 26th [pmk. AP26/ 76]

To G[eorge] J[ohn] ROMANES; Down, Beckenham.
AN on p.c.; 3 X 4 3/4; 1p. and add. [G. H. (sic) Romanes Esqr/ 18. Cornwall Terrace/ Regents Park/ London.N.W]

Cannot go to London today; has just received corr.'s note; will tell no one; hopes to be well enough to go to London tomorrow; will call upon corr.

489 [1876] Ap. 27th [pmk. AP27/ 76]

To G[eorge] J[ohn] ROMANES; 6. Queen Anne St. ("Down, Beckenham." is crossed out)
ANS on p.c.; 3 X 4 3/4; 1p. and add. [G. H. (sic) Romanes Esqr/ 18. Cornwall Terrace/ Regents Park.--]

Hopes to call upon corr. tomorrow morning.

490 [1876] Ap 29th

To [George John] ROMANES; 6. Queen Anne St[1]
ALS; 7 X $4\frac{1}{2}$; 1p.

Printed in full, with minor changes: Life of Romanes, 45.

1. Down address variant, type 8, is crossed out.

491 1876 May 3

To [August Wilhelm von HOFMANN]; Down (type 8)
LS; 8 X 5; 1p.

Thanks for corr.'s work on [Justus] Liebig; glad corr. has not forgotten CD.

492 1876 May 10.

To ?; Down (type 8) (last two lines are crossed out)
LS; 8 X 5; 2p.

Thanks for letter; will send sheets of new book [Cross and Self Fertilisation (1876)] before index is printed, as requested by corr.; this will not occur until October; is glad about newspaper and about twenty able collaborators.

493 [1876] May 29th.

To [George John] ROMANES; H. Wedgwoods Esq/ Hopedene/ Dorking[1]
ALS; 8 X 5; 6p.

First portion printed in full, with minor changes: *More Letters*, I, 363-65 (letter 275).
Next portion printed in full, with minor changes and with some overlap with previous portion: *Life of Romanes*, 49-50.
Small and redundant portion printed: *Life and Letters*, III, 204.

1. Down address variant, type 8, is crossed out.

494 [1876] June 4th

To [George John] ROMANES; Down (type 8)
ALS; 8 X 5; 4p.

Printed, with minor changes and minor omissions: *Life of Romanes*, 60-61.

495 [1876 ca. June 15][1]

To [George John] ROMANES; no location
A postscript, S by init.; 9 X $7\frac{1}{4}$; 4p., sketch

Romanes asks about Frank's [i.e. Francis Darwin's] discovery,[2] "but if I were like you in constant communication with spirits, I shd. not care even for a jelly-fish, & this is the strongest simile I can apply to you";[3] do not mention Frank's work until confirmed; Frank finds that leaves of teazle or *Dipsacus* have protoplasmic filaments extending from glands on footstalks; these filaments are stimulated by carbonate of ammonia or putrid meat and contract on exposure to alcohol; filaments appear to ingest matter; much work must still be done, and Frank is busy with another paper,[4] but "I can hardly doubt that we have here a highly organised plant which emits from its cells masses of protoplasm, identical with an amoeba or other Protozoan, for the sake of capturing particles of dead organic matter; & this seems to me a wonderful discovery"; letter by Dr. [Benjamin Ward] Richardson in *Nature* is capital.[5]

1. See note 5, below, for date.
2. See Francis Darwin, "On the Protrusion of Protoplasmic Filaments from the Glandular Hairs of the Common Teasel (*Dipsacus sylvestris*)," *Proc. R. Soc.*, *26* (1877): 4-8.
3. CD is joking about Romanes's interest in spiritualism; see *Life of Romanes*, 48-49.
4. "On the Glandular Bodies on *Acacia sphaerocephala* and *Cecropia peltata* Serving as Food for Ants...," *J. Linn. Soc.* (Botany), *15* (1875-1877): 398-409.
5. See Richardson, "Abstract Report to 'Nature' on Experimentation on

Animals for the Advance of Practical Medicine," Nature, Lond., 14 (1876): 149-52, 170-72, 197-99, 250-52, 289-91, 339-41, and 369-72. Published in the issue for June 15, 1876, and subsequent issues.

496 [?1876] June 30th

To ?; Down (type 8)
ALS; 8 X 5; 1p.

Has not heard of essay mentioned by corr.; would like copy if it is small and corr. has a spare one; will not answer it, as this takes time.

497 [1876] Sept. 15th [end. 1876/ 15 Septr.]

To [George Warde or Ebenezer] NORMAN; Down (type 8)
ALS; 8 X 5; 2p. and end. [1876/ 15 Septr./ Chas. Darwin/ on receipt/ of letter of/ condolence, for/ death of Daughter/ in Law--]

Thanks for note; it has been a terrible blow; [Amy Richenda Ruck Darwin] was sweet and gentle; son [Francis Darwin] has gone to North Wales where she was buried yesterday.

498 1876 Sep. 19

To [Peter Martin DUNCAN]; Down (type 8)
LS; 8 X 5; 2p.

While cleaning out a room, found a fossil coral and a long MS. account of it by [William] Lonsdale; these must have arrived while CD was ill, and they were forgotten; will send them to corr.

499 1876 Sep. 19

To [?William] MARSHALL; Down (type 8)
LS; 8 X 5; 2p.

Wishes to build a billiard room, with bedroom and dressing room above; wants corr. to undertake the project; come to Down; Horace [Darwin] does not know corr.'s London address; must hurry with project because winter is coming.

500 [1876][1] Sept. 29th

To [?William] MARSHALL; Down (type 8)
ALS; 8 X 5; 1p.

Please write to "Mr. Laslett/ Builder/ Farnborough/ Kent." and arrange to meet him at Down House on Tuesday; leaves home [for Leith Hill Place] on Wednesday [October 4], so cannot invite corr. to sleep at Down.[1]

1. See preceding letter, above, for year. For departure to Leith Hill, see "Darwin's Journal," 20. Laslett may have been William Emerson Laslett the solicitor.

501 [1876][1] Nov. 22d

To [?William] MARSHALL; Down (type 8)
ALS; 8 X 5; 6p.

Accepts Mr. Deards's estimate for the work, as his is cheaper; repay Laslett and [?Thomas Hayter] Lewis for expenses involved in estimates; make trap door to roof out of glass, and put flooring in attic crawl space, so lumber may be stored there; forget Horace [Darwin]'s idea about partition; wants to enlarge window in present hall; wants sealed airspace inside the outer walls.

1. See preceding letter, above, for date.

502 [1876]

To W[illiam Erasmus DARWIN]; no location (blue paper)
AL, S by init.; 8 X 5; 2p. (enclosure wanting)

Send enclosed to Capt. [?James Felix or ?Henry M.] Jones; proofs will reach corr. soon; new chapter seven will have to be much corrected by CD;[1] does not think much of [Edward Drinker] Cope's essay ["On the Origin of Genera," Proc. Acad. nat. Sci. Philad., 20 (1868): 242-300], read long ago when received; "He writes very obscurely, but is an excellent Naturalist. He looks, following [Jean Louis Rodolphe] Agassiz at a genus as something essentially distinct from a species, which I believe to be quite an error."

1. CD here refers to Fertilisation of Orchids (1877), which he was readying for publication in 1876. Thus the year for this letter. On the new chapter seven, see page vi of the book.

503 [1877] 1876 (sic) Jan 2.[1]

To [George John] ROMANES; Down (type 8)
LS;[2] 8 X 5; 3p.

Is pleased to propose corr. for Royal Society; will be at 2 Bryanston Street in London on the 6th; will sign certificate then; suggest others, not on the council [of the Royal Society], who might also sign it; [John Scott] Burdon-Sanderson should sign it; is reading corr.'s medusa paper;[3] will call on corr. early next week.

1. CD's amanuensis apparently wrote the wrong year. See subsequent letter, below, and check "Darwin's Journal," 20, to see that CD went to London of January 6, 1877, but not on January 6, 1876.
2. Francis Darwin, CD's amanuensis, penned a short note at end of letter.
3. "Further Observations on the Locomotor System of Medusae," Phil. Trans. R. Soc., 167 (1877): 659-752.

504 1877 Jan. 4.

To [George John] ROMANES; Down (type 8)
LS; 8 X 5; 1p.

From what corr. says, CD will not send certificate [proposing corr. for Royal Society]; will bring blank certificate to London; invites corr. to lunch at 2 Bryanston Street on Monday [January 8]; bring or send certificate.

505 1877 Jan. 10th [pmk. JA 10/ 77]

To Howard MILLER; Down (type 8)
ALS; 8 X 5; 1p. and env., add. [Howard Miller Esqr/ Elderton/ Armstrong Co./ Pa./ U. States]

Thanks for offer, which would be of value to anyone engaged in embryological studies, but CD is not so engaged.

506 1877 Feb. 15th-- [end. Feb. 16/ 77]

To R[?ichar]d [KE.?]; Down (type 8)[1]
ALS; 8 X 5; 2p., end. [Books sent & advised/ Feb. 16/ 77 Rd Ke.]

Please send Botanische Zeitung for 1857; send [Eugene] Michalet, ["Sur la Floraison des Viola de la Section Nomimium, de l'Oxalis acetosella et du Linaria spuria,"] Bull. Soc. bot. Fr., 7 [(1860):] 465[-70]; and send [Joseph] Duval-Jouve, ["Note sur Quelques Plantes Critiques du Flora monspeliensis de Linné,"] Bull. Soc. bot. Fr., 10 [(1863): 10-20]; send also H[enri Ernest] Baillon, "Sur l'Émission des Tubes Polliniques des Helianthemum," [Adansonia, Recueil Périodique d'Observations Botaniques, 2 (1861):] 56[-59]; send by rail.

1. Lower half of this printed address is highlighted with a drawn hand pointing to it.

507 1877 Feb. 25th

To Miss JACOBSON; Down (type 8)
AL in third person S; 6 X 5; 1p.

CD accedes to corr.'s request.

508 1877 Mar 26

To [George Henry Kendrick] THWAITES; Down (type 8)
LS; 8 X 5; 4p.

Thanks for specimens; sexual differences remind CD of complemental males in cirripedes; hopes [John Obadiah] Westwood will make a paper on subject;[1] wonders if specimens play role in fertilization of figs, a subject studied by Dr [Hermann] Crüger in West Indies; has seen account of corr. from Miss [Marianne] North;[2] has just examined Oxalis sensitiva specimen sent long ago, and to CD's surprise its flowers are trimorphic and cleistogamic.[3]

1. "Descriptions of the Insects Infesting the Seeds of Ficus Sycomorus and Carica," Trans. R. ent. Soc. Lond., 30 (1882): 47-60.
2. See entry for North in DNB; she was in Ceylon in 1876.
3. See Darwin, Different Forms of Flowers (1877), 322-24.

509 [1877][1] April 15th

To [George John] ROMANES; Down (type 8)
ALS; 8 X 5; 5p.

Is astonished that corr. was not on list [of newly-elected Fellows of Royal Society]; [Joseph Dalton Hooker], President [of the Royal Society] and others implied that age and position in scientific society (e.g. being a professor) weighed heavily, as did having been proposed many times; youth is a disqualification; Council [of Royal Society] acts honestly; corr.'s work is valuable; connot remember names of fifteen [new Fellows];[2] do not withdraw name; offers to re-propose corr. for election.

1. See letters to Romanes dated January 2 and January 4, 1877, above.
2. For names, see Proc. R. Soc., 26 (1877): 210.

510 1877 April 26th

To Otto ZACHARIAS; Down (type 8)
ALS; 8 X 5; 3p. and env., add. [Dr. Otto Zacharias/ Red: der Nordsee-Zeitung/ Geestemünde/ Germany]

Thanks for anomalous foot of the pig, but CD is not osteologist enough to pass judgment on it; is away from home; will send foot to Prof. [William Henry] Flower at Royal College of Surgeons, who has studied limbs of Ungulata and who is "a most careful & admirable observer"; will send notice if any "remarkable peculiarities" are found.

511 1877 May 3

From W[illiam] H[enry] FLOWER; Royal College of Surgeons of England./ Lincoln's Inn. Fields,/ (W.C.)
ALS; $7\frac{1}{4}$ X $4\frac{1}{2}$; 4p.

Thanks for pig's foot; it contains an extra digit, which may be either a reversion or a case of over-development; wants further information about other feet of affected pig and about feet of pig's relatives.[1]

1. See preceding letter, above.

512 1877 May 17th

From W[illiam] H[enry] FLOWER; Royal College of Surgeons of England./ Lincoln's Inn Fields,/ (W.C.)
ALS; $7\frac{1}{4}$ X $4\frac{1}{2}$; 3p., sketch

Thanks for information about pig's foot in letter of May 10; sends catalog of their teratological series; mentions another pig foot specimen with similar deformity, specimen number 297A; glad corr. will collect similar cases.[1]

1. See preceding letter, above. This letter may be to Otto Zacharias, and not to CD, but there is no indication of such on original.

513 1877 May 23d.

To [George John] ROMANES; Down (type 8)
ALS; 8 X 5; 4p.

Thank [Charles] Grant [Blairfindie] Allen for having sent his book [Physiological Aesthetics (London: H. S. King, 1877)]; thanks for letter; "good Heavens what a lot of books there is to read"; had not heard of "rub"--presumably over spontaneous generation--at Royal Society;[1] [Thomas Henry] Huxley said [John] Tyndall's work about old germs withstanding boiling was decisive;[2] would like to see question settled; Frank's [i.e. Francis Darwin's] paper on teazle has been rejected, to CD's chagrin;[3] none of sons can attend corr.'s lecture;[4] corr.'s negative results concerning spiritualism please CD, who thinks Williams [the spiritualist] is "a very clever rogue".

1. See Glenn Vandervliet, Microbiology and the Spontaneous Generation Debate during the 1870's (Lawrence, Kansas: Coronado Press, 1971).
2. See Tyndall, "Further Researches on the Deportment and Vital Resistance of Putrefactive and Infective Organisms...," Proc. R. Soc., 26 (1877): 228-38; and idem, "Further Researches on the Deportment and Vital Persistence of Putrefactive and Infective Organisms...," Phil. Trans. R. Soc., 167 (1877): 149-206.
3. See Francis Darwin, "On the Protrusion of Protoplasmic Filaments

from the Glandular Hairs of the Common Teasel (Dipsacus sylvestris)," Q. Jl microsc. Sci., 17 (1877): 169-74 and 245-72.
4. Romanes, "Evolution of Nerves and Nervo-Systems," Not. Proc. Meet. Memb. R. Instn, 8 (1875-1878): 427-48.

514 [1877][1] May 27th

To [George John] ROMANES; Down (type 8)
ALS; 8 X 5; 4p.

Thanks for letter; does not know what referees said about Frank's [i.e. Francis Darwin's] paper;[1] [Edwin] Ray Lankester and [John Hutton] Balfour have seen [teasel] filaments; latter is sure they are protoplasm; "It is terribly discouraging to a young worker to have his work rejected because the facts are quite new";[1] will show corr.'s letter to Frank; will read corr.'s lecture when it appears in Nature;[2] deception during seances is wicked and scandalous; will take a month's rest soon [from June 8 to July 4, 1877];[3] has read a third of [Charles] Grant [Blairfindie] Allen's book, finding it clever but too "deductive"; Allen neglects effects of habit.[1]

1. See preceding letter, above.
2. "Evolution of Nerves and Nervo-Systems," Nature, Lond., 16 (1877): 231-33, 269-71, and 289-93.
3. See "Darwin's Journal," 20.

515 1877 June 5.

To [George John] ROMANES; Down (type 8)
LS; 8 X 5; 2p.

Thought corr. would like the following extract, found by chance, and bearing on effects of habit in passage of nervous force.[1]

1. See Jean Lamarck, Philosophie Zoologique, 2v. (Paris: Baillière, 1830), II, 318, paragraph beginning: "Dans toute action, le fluide...."

516 [1877] June 11th.

To [George John] ROMANES; "My address will be/ 'Bassett, Southampton.'"[1]
ALS; 8 X 5; 4p.

Printed in full, with minor changes: More Letters, I, 367-68 (letter 279). p. 367, line 7, change "purchases" to "preserves".

1. Down address variant, type 8, is crossed out. Also crossed out is "Leith Hill Place".

517 [1877 July 16th][1]

To [Rudolf August Birminghold LUDWIG][1]; Down (type 8)
LS; 8 X 5; 2p.

Received letter and essay dedicated to CD, dated July 4, only today;[2] it is interesting to discover in ancient forms any characters now present in distinct species or subgroups, and Crocodilus Darwinii is excellent illustration of this.

1. Pencil note on third page reads: "addressed Herrn Rudolphe Ludwig/ Waldstrasse 49/ Darmstadt/ July 16th 1877".
2. See Rudolf Ludwig, Fossile Crocodiliden aus der Tertiärformation des Mainzer Beckens... (Cassell, 1877).

518 [1877] Augt. 9th

To [George John] ROMANES; Down (type 8)
ALS; 8 X 5; 6p.

Printed in full, with minor changes: Life of Romanes, 56-57. p. 57, line 4, change "half of the botanists" to "half-a-score of Botanists". Also printed, with changes: More Letters, II, 406-07 (letter 731).

519 [1877] Aug. 10th

To [George John] ROMANES; Down (type 8)
ALS; 8 X 5; 3p.

Printed in full, with minor changes: Life of Romanes, 62-63. Printed, with minor changes and minor omissions: More Letters, II, 407-08 (letter 733). p. 407, line 1, change "went" to "wrote".

520 1877 Sept. 15

From Léo ERRERA; Venise.
Draft of ALS; 8 X 5½; 3p. (in French)[1]

Has studied heterostylism for some years in the Primula elatior; has prepared manuscript on subject with G[ustave] Gevaert, in which there is a correction of a point made in Darwin, Cross and Self Fertilisation (1876); wishes to send portion of manuscript that deals with CD's book to CD for review; reply in English.

1. Translation into English provided by editor's wife, Nancy E. Carroll.

521 1877 Sep 18. [pmk. SP18/ 77]

To Léo ⌈Abram⌉ ERRERA; Down (type 8)
LS; 8 X 5; 2p. and env., add. ⌈Mons: Léo Errera/ 6A. Rue Royale 6A/ Brussels/ Belgium⌉

Would be happy to look at corr.'s MS.; to save CD's time, mark pages which CD should scan; will read paper in entirety when published;[1] has recently published Different Forms of Flowers ⌈(1877)⌉, in which results of some experiments on P. elatior are given.[2]

1. See Errera and Gevaert, "Sur la Structure et les Modes de Fecondation des Fleurs et en Particulier sur l'Hétérostylie du Primula elatior," Belg. Soc. Bot. Bull., 17 (1878): 38-181 and 182-248.
2. See preceding letter, above.

522 1877 Sept. 30

From L⌈éo Abram⌉ E⌈RRERA⌉; Bruxelles/ 6A rue royale.
Draft of ALS; various sizes; 3p. (in French)[1]

Thanks for prompt letter; has marked the essential passages of the MS. for CD's comment in blue crayon; have borrowed from ⌈A. von Marilaun⌉ Kerner the terms "autogamy", "geitonogamy", "xenogamy", and "allogamy"; discusses the advantages of these; points out the main elements of the argument in the MS.; thanks CD for copy of Darwin, Different Forms of Flowers ⌈(1877)⌉; mentions terminology used by Herman Müller and ⌈F.⌉ Delpino.[1]

1. Free translation into English provided by editor's wife, Nancy E. Carroll. See preceding letter, above.

523 1877 Oct 4th ⌈pmk. OC4/ 77⌉

To Léo ⌈Abram⌉ ERRERA; Down (type 8) (last two lines crossed out)
LS; 8 X 5; 4p. and env., add. ⌈Monsieur Léo Errera/ 6A Rue Royale/ Brussells⌉

Printed in full, with minor changes: More Letters, II, 413-14 (letter 741).

524 1877. Oct 10

From L⌈éo Abram⌉ E⌈RRERA⌉; no location
Draft of ALS; 8½ X 6½; 2p. (in French)[1]

Thanks for aid; has already begun work upon inconspicuous flowers; explains why he concluded that CD believed that autogamy generally was more advantageous than geitonogamy.[1]

1. Translation into English provided by editor's wife, Nancy E. Carroll. See also preceding four letters, above.

525 1877 Oct 15.

To ?; Down (type 8)
LS; 8 X 5; 2p.

Thanks for curious photographs; will read account of them with interest; appearance of same abnormality in different races of humans in different environments has always seemed remarkable to CD; did not know corr.'s pupil R[udolf] von Willemoes-Suhm, but had read his papers, esp. on development of cirripedes; had "highest opinion of his abilities. His death has been a very serious loss to natural science."

526 [1877 December 1-2][1] Saturday night, Sunday morning

To [George John] ROMANES; Down (type 8)
ALS; 8 X 5; 4p.

Printed in full, with minor changes: Life of Romanes, 66-67.
Printed, with minor changes and minor omissions: More Letters, I, 370-71 (letter 281).

1. Date determined from date of Romanes's reply; see Life of Romanes, 68-69.

527 1877 Dec 5.

To [George John] ROMANES; Down (type 8)
LS;[1] 8 X 5; 3p.

CD's gardener does not know what onion worm is; onions can be planted at Down in part of garden where onions have not grown for 30 years; will take charge of experimental plants for corr.; thanks for letter; Romanes overestimates CD and his works; it would be absurd for CD to suddenly be called "Dr".

1. Francis Darwin, CD's amanuensis, penned a short note at end of letter, recommending that Romanes try "Thymol" upon his medusas.

528 1877 Dec. 11

To ?; Down (type 8)
LS; 8 X 5; 2p.

Thanks for letter; would like to see six letters of grandfather [Erasmus Darwin] published, but grandfather's name and that of [Johann Albrecht Heinrich] Reimarus are so unknown that no editor would want to publish the letters; would like to see letters if brought to England.

529 1877 Dec 27. [pmk. APR/ 27]

To R[euben A.] BLAIR; Down (type 8)
LS; 8 X 5; 2p. and env., add. [R. Blair Esq/ Sedalia/ Missouri/ U.S.A.]

Thanks for news of the case of the goose; seems to show inheritance of the effects of injury; such cases are rare and should be examined; send wing to CD, if possible; see if goose had normal offspring before injury.

530 [1877-1878][1] Jan. 24th

To [George John] ROMANES; Down (type 8)
ALS; 8 X 5; 6p.

Printed in full, with minor changes: *Life of Romanes*, 105-07.

1. CD mentions "perforation of the corolla by bees" in this letter as being on page 425 of Darwin, *Cross and Self Fertilisation*. This only occurred in the first edition, published November 10, 1876; see *Freeman*, page 31. Therefore, 1877 is lower endpoint. A new, second edition appeared in 1878 without any mention of perforation on page 425; this sets upper endpoint for letter.

531 1878 March 7.

To [George John] ROMANES; Down (type 8)
LS; 8 X 5; 2p.

CD's gardener says onions sown before middle of March may be injured by frost, but if CD delays in planting, onions will not be ready by May; should CD take risk and plant now; is glad corr.'s essay will appear in the Fortnightly; was thinking about it the other day; books were sent by rail in morning.[1]

1. See Romanes, "The Scientific Evidence of Organic Evolution: A Discourse," *Fortnightly Rev.*, *30* (1881): 739-58.

532 1878 March 28th

To ?; Down (type 8)
ALS; 8 X 5; 1p.

Thanks for letter; temperament or mental disposition depends on structure of the brain, so nature of milk digested probably does not affect character of child; case of diseases is different; facts about starling imitating cries and songs of other birds are curious, but "I have become so absorbed with vegetable physiology that it is not likely I shall ever again attend to the minds of the higher animals."

533 [1878] April 9th

To [George John] ROMANES; Down (type 8)
ALS; 8 X 5; 3p.

Condolences [on loss of sister];[1] leaves home on 27th and returns on May 13; invites corr. to Down before or after these dates; onions are not yet up; doubts that corr. could do anything with onions before leaving for Scotland.[2]

1. See Life of Romanes, 69-71.
2. On onions, see also letter to Romanes, March 7, 1878, above.

534 1878 April 12

From W[illiam] H[enry] FLOWER; Royal College of Surgeons of England/ Lincoln's Inn Fields/ (W.C.)
ALS; 8 X 5; 4p.

Was delayed in looking at goose's wing; sketches the wing's bones on the back of [Reuben A.] Blair's letter; bones are normal, but appear to have been abnormally flexed; proposes "far-fetched" hypothesis in which such an injury might have led to "some disturbance in the functons of the nerves which control the muscles of the same region" in the offspring; must examine several wings, especially of very young birds, to check this; wing of original goose should be examined as well; case is worth following up.[1]

1. See also letter to Blair, December 27, 1877, above. With this letter is a copy of same, plus an env., add. to "R. A. Blair Esq/ Sedalia/ Missouri/ U.S.A" and pmk. "Beckenham/ DE27/ 77". It should accompany earlier letter, of course.

535 1878 Ap. 14.

To [Reuben A. BLAIR]; Down (type 8)
LS; 8 X 5; 4p. and photo of goose (enclosure wanting)[1]

Thanks for wing of goose, photograph, and letter; enclosed explains delay of [William Henry] Flower; send wings requested by Flower directly to Flower; CD will pay expenses for this undertaking; return letters to CD; did old gander bleed when it was wounded; did wound suppurate; did wing heal quickly or slowly; these questions relate to inheritance of mutilations.

1. Enclosure is obviously Flower's report, which is abstracted above as a letter to CD.

536 [1878 May 23] [end. 1878; pmk. MY23/ 78]

To Rob[er]t THOMSON; no location
ALS; 4½ X 7 3/4; 1p. and env., add. [Robt Thomson Esq/ Beaufort/ South Carolina/ U.S.A.], end. [Charly Darwin/ 1878]

Orders "large pot of *soft* spermaceti ointment".

537 [1878][1] June 16th

To [George John] ROMANES; Down (type 8)
LS; 8 X 5; 4p. (enclosures wanting)

Wrote notes on insects 40 years ago and used them in long chapter on instinct written four or five years before Darwin, *Origin* [(1859)]; sends two pages of this chapter; offers entire chapter of 110 pages for Romanes's use, but presumes it is too late to be of any good; Frank [i.e. Francis Darwin] took the baby [i.e. Bernard Richard Meirion Darwin] to see train and called engine a "puff puff"; baby now calls all vehicles and chimneys and fires (lighted or unlighted) by the name "boo boo"; more onions from last lot have died, apparently because of wet weather, but total deaths are few;[1] one pea has sent up an aborted shoot of a half plumule.

1. See letter to Romanes, April 9, 1878, above.

538 [1878] June 19th

To [George John] ROMANES; Down (type 8)
ALS; 8 X 5; 3p.

Printed in full, with minor changes: *Life of Romanes*, 72.

539 1878 July 1.

To [a son of Edward TRUELOVE]; Down (type 8)
ALS; 8 X 5; 3p. and end. [Chas. Darwin]

Private letter; had not heard of corr.'s father until his trial; cannot sign memorial for this reason; has no reason to doubt corr.'s father's character; "I have not seen the Moral Phys. by R[obert] D[ale] Owen, but I have heard the nature of its contents; & though I am strongly opposed to all such views & plans, yet the Conviction of your Father seemed to me very harsh, as the publication of works of this nature can hardly be considered as obscene in the ordinary sense of the word."[1]

1. See: Owen, *Moral Physiology; or, a Brief and Plain Treatise on the Population Question...* (New York: 1831); and Edward Truelove, *In the High Court of Justice. Queen's Bench Division, February 1, 1878. The Queen v. E. Truelove, for Publishing the Hon. R. D. Owen's "Moral Physiology"...* (London: 1878). Copy of the latter is at British Museum.

540 1878 July 4th

To [Reuben A. BLAIR]; Down (type 9)
ALS; 8 X 5; 1p.

Thanks for letter and for kindness; has written to [William Henry] Flower with all particulars and has urged Flower to examine wings carefully; there may be delay caused by holiday; corr. will receive word.[1]

1. See letter to Blair, April 14, 1878, above.

541 [1878] July 16th [pmk. JY16/ 78]

To F. G. STOKES; Down (type 8)
ALS; 8 X 5; 2p. and env., add. [F. G. Stokes Esq/ The Nook/ Alma Road/ Windsor], end. [C. Darwin/ 2762]

Corr. will find interesting points to observe; if corr. has musical ear, he should study the various intonations which children just before and just after learning to speak; these sounds are "highly expressive".

542 1878 July 17.

From R[euben] A. BLAIR; Sedalia Mo.
ALS; 10 X 8; 4p.

Thanks for letter of April 14; has sent pair of ten-month-old wings, pair of ten-day-old wings, one three-month-old gander's wing, and the old gander's wing, to [William Henry] Flower; had difficulties obtaining wings; abnormality did not appear in offspring until fifth week; parents of the old goose and gander had no defects; injury did not produce blood or suppuration; speculates about nature of injury.[1]

1. See letter to Blair, July 4, 1878, above.

543 1878 Aug. [4? 11?]

To [William Henry] FLOWER; Down (type 8)
ALS; 8 X 5; 4p. (enclosure wanting)[1]

Has heard from R[euben] A. Blair, who has sent specimens as described in enclosed letter; since [Charles Edouard] Brown-Sequard's observations, all cases of inheritance from injury seem worthy of study; encloses Blair's letters to CD; encloses Flower's note to CD on subject; please tell CD of results, so that Blair can be informed; man in Leeds offered CD a pigeon born without eyes and without sockets in skull; told man to send it to Flower.[1]

1. See previous letter, above.

544 [1878] Aug 8th [pmk. DORKING/ AU8/ 78]

To [Léo Abram ERRERA]; Leith Hill Place/ Dorking[1]
ALS; 8 X 5; 1p. and env., add. [A/ Mon. Léo Errera/ Morley's Hotel/ Trafalgar Sqr/ London. and 152 (clique?) de Mer/ Blankenberghe/ Belgium], end. [letter of/ Aug 8th/ (1878)]

Sorry corr. went to Down for nothing; left home before the post arrived yesterday; hopes corr.'s researches on cross-fertilization progress well.

1. Down address variant, type 8, is crossed out.

545 1878 August 12

From [Léo Abram ERRERA]; Bkbgh [i.e. Blankenberghe]
Retained copy of AN; $5\frac{1}{4}$ X 2 3/4; 2p.

Thanks for letter; hopes to find CD at home some other time; invites CD to call upon Errera if CD ever comes to the Continent; congratulations on election to French Academy of Sciences.[1]

1. See preceding letter, above.

546 1878 Aug. 20th/ 28th

To [George John] ROMANES; Down (type 8)
ALS; 8 X 5; 8p.

Printed in full, with minor changes: Life of Romanes, 74-77.
Also printed: Life and Letters, III, 225; and More Letters, II, 48-49 (letter 417).

547 [1878] Sept. 2d

To [George John] ROMANES; Down (type 8)
ALS; 8 X 5; 3p. (enclosure wanting)

Printed, with minor changes: Life of Romanes, 78. At end of letter is: guesses that Lady [Mary Farrer] Hobhouse is trustworthy;[1] sends [Joseph Rémi Léopold] Delboeuf, [?La Psychologie comme Science Naturelle, Son Présent et Son Avenir (Brussels: 1876)];[2] enclosed is worth getting; mentioned first edition to corr.; it is smallish book; "He is a Spencerian to the back-bone."[3]

1. See Life of Romanes, 76; and More Letters, II, 49 (letter 417).
2. See More Letters, II, 48 (letter 417); and Life of Romanes, 75-76.
3. Probably a reference to Alfred Victor Espinas, Des Sociétés Animales, 2nd ed. (Paris: G. Baillière, 1878). See Life of Romanes, 79.

548 [1878][1] Sept 21st

To [George John] ROMANES; Down (type 8)
ALS; 8 X 5; 2p.

Order copy of last Thursday's Spiritualist;[1] it contains exposure of Williams [the medium];[2] H[ensleigh] Wedgwood brought it to Down; Wedgwood admits that Williams is a rogue, but believes him to be a medium nonetheless; Wedgwood claims further to have seen John King, notwithstanding the old beard and dirty ghost clothes; this is a psychological curiosity![3]

1. See "An Exposure of Imposture," The Spiritualist Newspaper, A Record of the Progress of the Science and Ethics of Spiritualism, 13, September 20, 1878. Year for this letter determined by this publication date.
2. See also letter to Romanes, May 23, 1877, above.
3. On this whole subject, see Edward Clodd, The Question: 'If a Man Die, Shall He Live Again?', A Brief History and Examination of Modern Spiritualism (London: Grant Richards Ltd., 1917), passim, esp. 128-29.

549 1878 Sept. 28th

To ?; Down (type 8)
LS; 7 X 4½; 1p. (enclosure wanting)

Private letter; encloses copy of the Medium with an exposure of Williams the spiritualist; hopes corr. will publish an account of the exposure.[1]

1. See preceding letter, above.

550 [?1878][1] Oct 1st

To [George John] ROMANES; Down (type 8)
ALS; 8 X 5; 1p.

Has just read corr.'s article in Fortnightly;[1] thinks it good; "I honour you for taking so much trouble pro bono publico of the mass of readers"; article in the Edinburgh must be nonsense; editor should not have admitted such an article; "But the poor old Edinburgh seems to have played out its part & is ready for extinction."[2]

1. See, perhaps, Romanes, "The Beginning of Nerves in the Animal Kingdom," Fortnightly Review, 30 (1878): 509-26. Year for letter determined by this publication date.
2. See the review of geology texts in Edinb. Rev., 147 (Jan.-April, 1878): 183-200. This reference suggested by editor's wife, Nancy E. Carroll.

551 1878 Nov 6th

From W[illiam] H[enry] FLOWER; Royal College of Surgeons of England./ Lincoln's Inn Fields,/ (W.C.)
ALS; $7\frac{1}{2}$ X $4\frac{1}{2}$; 7p. and enclosure[1]

Flower and assistant Dr. Larson have dissected wings sent by [Reuben A.] Blair; sends Larson's report;[1] will give summary of it here; deformity is same in all young birds and resembles clubfoot in man; wing of old gander, however, is "perfectly normal" and has never been broken or dislocated; concludes that this is merely a case of mutual deformity in siblings; does not know if this is common, but saw a similar specimen during summer; returns letters.[2]

1. Enclosure is Larson's report; see appendix, below.
2. See letter to Flower, Aug. [4? 11?], 1878, above.

552 [1878 November] 20th Wednesday [pmk. NO20/ 78]

To G[eorge] J[ohn] ROMANES; at R. B. Litchfield/ 4, Bryanston Street/ Portman Square, W.
ANS on p.c.; 3 X 4 3/4; 1p. and add. [G. J. Romanes Esqr/ 18. Cornwall Terrace/ Regents Park]

Will call upon corr. tomorrow morning; send card if corr. will not be home then.

553 [1878] Dec. 5th

To [George John] ROMANES; Down (type 9)
ALS; 8 X 5; 4p.

Printed in full, with minor changes: <u>Life of Romanes</u>, 85-86.

554 1878 Dec. 9th.

To [Reuben A. BLAIR]; Down (type 9) (last two lines crossed out)
ALS; 8 X 5; 3p. and enclosures[1]

[William Henry] Flower has been ill, so CD's reply has been delayed; encloses Flower's report and letter;[1] there is no connection between the deformity and the injury; it must be a case of "post hoc" and not "propter hoc"; "I am well accustomed in my experimental work to get definite result only once in three or four times, & thus alone can science prosper."

1. Enclosures are listed separately elsewhere in calendar; see letter from Flower, November 6, 1878, above.

555 [1878] Dec. 11th

To /̄George John_7 ROMANES; Down (type 8)
ALS; 8 X 5; 1p.

Thanks for letter;[1] has just received "Vegetative Bastarderzeugung durch Impfung" by /̄Hugo_7 Lindemuth, an agriculturist and gardener;[2] it has three plates, and CD offers it to corr.; CD has not received papers which corr. says he has sent.

1. See Life of Romanes, 86-88.
2. See Lindemuth, "Ueber Vegetative Bastarderzeugung durch Impfung," Landw. Jbr, 7 (1878): 887-939.

556 /̄1878_7 Dec. 27th /̄pmk. DE28/ 78_7

To G/̄eorge_7 J/̄ohn_7 ROMANES; Down, Beckenham.
AN on p.c.; 3 X 4 3/4; 1p. and add. /̄G. J. Romanes Esqr/ 18. Cornwall Terrace/ Regents Park/ London N.W._7

Read Prof. /̄William_7 James, "Brute and Human Intellect," Journal of Speculative Philosophy, /̄12_7 (1878): 236-/̄76_7.

557 1878 Dec. 31

To /̄William Rathbone_7 GREG; Down/ Beckenham
LS; 10¼ X 8; 5p.

Has read chapter on design and thinks it clever; corr.'s son should read book by "Physicus" and published by Trübner; none of corr.'s son's objections are new; see /̄Darwin_7, Variation under Domestication /̄(1868)_7, II, last three pages, on design in variation; /̄Gaston de_7 Saporta, Geinitz and Oppell, and /̄Melchior_7 Newmayr have found close series of fossil forms; fossil bones sometimes do not reveal fine differences; for CD's views on crossing eliminating new variations, see Variation under Domestication, chap. XX; corr.'s son forgets effects of isolation on crossing; corr.'s son "says I rest exclusively on natural selection; whereas no one else as far as I know has made so many observations on the effects of use & disuse. Nor do I deny the direct effect of external conditions, tho' I probably underrated their power in the earlier editions of the Origin. We know far too little about the laws of inheritance to argue about them: what is certain is that some new variations are strongly inherited from the first, whilst some seem never to very rarely to be transmitted; nor until trial is made can we predict what will be the result"; /̄Richard_7 Owen says open spaces in teeth of race horses were specially designed for bit; "no doubt the little jockeys were specially designed to ride them"; dismisses final cause for lowering of specific gravity of water upon freezing, giving bismuth as counterexample.

558 /̄1879 January 18_7 /̄pmk. JA18/ 79_7

To G/~eorge_7 J/~ohn_7 ROMANES; Down, Beckenham.
N on p.c.; 3 X 4 3/4; 1p. and add. /~G. J. Romanes Esq/ 18 Cornwall Terrace/ Regents Park/ London NW_7

Sees advertised /~Tito_7 Vignoli, /~Über das_7 Fundamentalgesetz der Intelligenz im Thierreich/~e (Leipzig: 1879)_7.

559 1879 Jan 31.

To ?; Down (type 8)
LS; 8 X 5; 2p.

Will not sign corr.'s petition because does not consider himself as "coming with the class" who will sign it; thinks establishment of a professorship of pathology at Cambridge would be desirable; one of CD's sons says corr. errs about abolition of medical fellowships at Trinity.

560 1879 Feb. 25th /~pmk. FE25/ 79_7

To Karl HÖCHBERG; Down (type 9) (last two lines crossed out)
ALS; 8 X 5; 3p. and env., add. /~Herrn Karl Hochberg/ Castagnola/ presso Lugano/ Switzerland_7

Has never attended to vegetarian diet; only valuable evidence would be statistics on countries where inhabitants live on widely different diets; hardest workers CD ever saw, the miners in Chile, lived exclusively on vegetable diet including seeds of Leguminosae; but Gauchos are active and live almost exclusively on meat; in tropical Africa there is an extraordinary craving for meat at intervals, but inhabitants eat largely of seeds of Leguminosae, and Arachis hypogaea is widely cultivated.

561 1879 March 5th /~pmk. MR 5/ 79_7

To Léo /~Abram_7 ERRERA; Down (type 9)
ALS; 8 X 5; 2p. and env., add. /~A/ M. Léo Errera/ aux soins G. Magolez/ Libraire Rue de l'Impératrice/ Bruxelles._7

Has just returned from a week's absence to find present from corr. and /~Gustave_7 Gevaert;[1] will read it soon; corr.'s work is more than a vulgarization or popularization.

1. Errera and Gevaert, "Sur la Structure et les Modes de Fecondation des Fleurs et en Particulier sur l'Hétérostylie du Primula elatior," Belg. Soc. Bot. Bull., 17 (1878): 38-181 and 182-248. See also letter to Errera, September 18, 1877, above.

562 1879 March 13th [pmk. MR13/ 79]

To Léo [Abram] ERRERA; Down (type 9) (last two lines crossed out)
ALS; 8 X 5; 3p. and env., add. [Monr. Léo Errera/ aux soins G. Magolez/ Libraire - Edituer/ Rue de l'Impératrice/ Bruxelles/ Belgium.]

Has read corr.'s book, containing new, original, and ingenious views;[1] likes case of the staminoid [sic; staminode] of Pentstemon and observations on insects rarely visiting some varieties; that corr. and [Gustave] Gevaert will continue labors is "the earnest wish of one who is too old to do much more himself"; engraving of Primula is good; would like photograph of corr.

1. See preceding letter, above.

563 [ca. 1879 March 13][1]

From [Léo Abram ERRERA]; no location
Retained copy of AL, S by init.; 8 3/4 X 5½; 2p.

Has delayed answering until a photograph of Errera was available; wishes photograph of CD in return; corr. and [Gustave] Gevaert are encouraged by CD's praise.

1. This is obviously a reply to the preceding letter, above.

564 1879 June 4th

To ?; Down (type 8)
ALS; 8 X 5; 1p.

Corr.'s note has much on measurement about which CD is incompetent; has asked son-in-law, [Richard Buckley] Litchfield, to look into the case and report to corr.

565 1879 July 9th

To [Edwin Ray] LANKESTER; Down (type 8)
ALS; 8 X 5; 3p.

Thank authoress of "Key-notes" for her present; has lost appreciation for poetry, however; is glad corr. is to spend more time on original investigations, because corr. does "splendid work"; leaves for holiday in lakes soon; will read poetry there.

566 1879 July 23d

To ⌈George John⌉ ROMANES; Down (type 8)
ALS; 8 X 5; 3p. (enclosure wanting)

Is grieved at what corr. says about ⌈Charles⌉ Grant ⌈Blairfindie⌉ Allen; sends check for £25 for the subscription; will give more if needed; sorry corr. and wife ⌈Ethel Duncan Romanes⌉ could not come to Down; has been ill; leaves for Coniston for a month on August 1; hopes weather stays as good as it is; encloses paper, but ⌈Thomas⌉ Meehan is a very inaccurate observer; good luck with medusas; regards to wife.

567 ⌈1879⌉[1] Sept. 14th

To ⌈George John⌉ ROMANES; Down (type 8)
ALS; 8 X 5; 4p.

Printed, with minor changes and minor omissions: Life of Romanes, 97-98. At end of letter is: how is poor ⌈Charles⌉ Grant ⌈Blairfindie⌉ Allen?

1. Year determined by reference to Coniston in postscript; see "Darwin's Journal," 21. See also preceding letter, above.

568 1879 Dec. 4th

To ?; Down (type 8)
ALS; 8 X 5; 1p.

Thanks for note; remembers having seen same appearance in horses, but is unlikely to write ever again about variation of domesticated animals.

569 ⌈?1879-1882⌉[1] 25th

To ⌈George John⌉ R⌈OMANES⌉; Down
ANS; 6 3/4 X 5¼; 1p. (enclosure wanting)

Enclosed received this morning; they may be worth taking to Linnean Society; agrees about the ⌈Charles⌉ Grant ⌈Blairfindie⌉ Allen affair.[1]

1. The Grant Allen episode began with letter to Romanes, July 23, 1879, above; CD died in 1882.

570 1880 Feb 3.

To ⌈George John⌉ ROMANES; Down (type 8)
LS; 8 X 5; 3p.

Printed, with minor changes: Life of Romanes, 94. line 6, add: two enclosed letters from [?John Fletcher] Moulton on the subject are worth reading, but they are private; please return them; Moulton does not think much of [Frederick] Guthrie as a physicist. At end of letter is: is glad to read [Charles] Grant [Blairfindie] Allen's letter; "we have all done a good work."[1]

1. See preceding letter, above; for reply, see Life of Romanes, 95-96.

571 1880 Feb 5.

To [George John] ROMANES; Down (type 8)
LS; 8 X 5; 4p. (enclosure wanting)

Printed, with minor changes and minor omissions: More Letters, II, 51-52 (letter 420). At end of letter is: returns diagram.

572 1880. Apr. 16th [pmk. AP16/ 80]

To W. TEARLE; Down (type 8)
ALS; 8 X 5; 2p. and env., add. [W. Tearle Esqr/ Cambridge St/ St Neots/ Hunts.]

Private letter; cannot aid corr.; any remarks CD could make would add to corr.'s doubts on subjects which corr. considers sacred; "...every man ought to weigh for himself impartially & anxiously all the arguments for & against any revelation ever having been made to man.--"

573 1880 May 5th

To Cousin [Hensleigh WEDGWOOD]; Down (type 9)
ALS; 8 X 5; 3p.

CD and corr. have not met in some time; no good scientific journal would publish corr.'s essay; "Science progresses only by the discovery of new facts & direct deductions from them"; there have been too many attempts to reconcile Genesis and Science; considers death of corr.'s brother [Josiah Wedgwood, of Leith Hill Place] a grievous loss; there never existed "a man with a sweeter disposition."

574 [1880] Nov. 14th

To [George John] ROMANES; Down (type 9)
ALS; 8 X 5; 4p.

Printed in full, with minor changes: Life of Romanes, 101-02.

Figure 8. Charles Robert Darwin. Bas-relief executed after Darwin's death by Thomas Woolner (1825-1892) and cast by Wedgwood. A duplicate hangs in Darwin's room at Christ's College, Cambridge. (American Philosophical Society)

575 1880 Nov. 15th

To [George John] ROMANES; Down (type 8)
ALS; 8 X 5; 4p.

Printed in full, with minor changes: Life of Romanes, 102-03.

576 1880 Dec 20th

To [George John] ROMANES; Down (type 9)
ALS; 8 X 5; 4p.

Thanks for pleasing note;[1] Frank [i.e. Francis Darwin] thought that his papers interested no one;[2] on corr.'s experiments, suggests canary grass and cabbage seedlings when short; fill pots to top, so rims will not make shade; plants should be kept in complete darkness until time for experiment, making them sensitive to light and upright; comparing effects of intermittent light upon plants with effects upon animals will be interesting; corr.'s last sentence amused CD; [Ethel Duncan] Romanes is right to forbid monkey from nursery.

1. See Life of Romanes, 104-05.
2. See Francis Darwin, "The Theory of the Growth of Cuttings...," J. Linn. Soc. (Botany), 18 (1881): 406-19; and idem, "On the Power Possessed by Leaves of Placing Themselves at Right Angles to the Direction of Incident Light," ibid., 420-55.

577 [1880] Dec. 25th

To [John Maurice HERBERT]; Down (type 9)
ALS; 8 X 5; 4p.

Printed in full, with minor changes: Sir Gavin de Beer, ed., "The Darwin Letters at Shrewsbury School," Notes Rec. R. Soc. Lond., 23 (1968): 73-74. p. 73, line 4, change "also like" to "also much like" and change "something" to "anything". p. 73, line 12, change "me" to "us".

578 [1880-1881][1]

To [George John] ROMANES; 4 Bryanston St.[1]
ALS; 8 X 5; 1p.

Thanks; if able to see [Thomas Henry] Huxley, will not be able to see corr. before noon; will see corr. in afternoon if corr. calls then.

1. Down address variant, type 8, is crossed out. This letter is in

the shaky hand characteristic of CD from late 1880 onwards; CD's last visit to Bryanston Street was in 1881.

579 1881 Jan 20th

To ?; Down (type 9)
ALS; 8 X 5; 1p.

Required address is "A. R. Wallace Esq./ Pen-y-Bryn/ St. Peters Road/ Croydon."; letter with unexpected news about date of pension [for Alfred Russel Wallace] makes CD happy, because Wallace can rest immediately from literary work and devote strength to natural science; receipt of letters has been delayed by snow drifts.

580 1881 Jan 24

To [George John] ROMANES; Down (type 8)
LS; 8 X 5; 3p. (enclosure wanting)

Romanes can buy young pigs, but they must be inspected first and none will pass inspection because infection is about; another obstacle would be the immense snow drifts; keep enclosed letter from [Jean Henri] Fabre.[1]

1. For CD's reply to Fabre's letter, see *More Letters*, I, 385-86 (letter 298).

581 1881 Jan 28th

To [George John] ROMANES; Down (type 9)
ALS; 8 X 5; 4p.

Has just read corr.'s review in *Nature*; [Samuel Butler] will "smart under your stricture" unless he is "as callous as a rhinoceros"; corr. has been too severe; corr. is right to attribute [Butler's] conduct to "the disappointment of his inordinate vanity"; it is psychological curiosity that [Butler] should suppose that [Thomas Henry] Huxley lectured and [John] Murray advertised [Ernst Krause, *Erasmus Darwin* (1879)] because of his book [Butler, *Evolution Old and New...* (London: Hardwicke & Bogue, 1879)]; [Butler] will hate corr.; thanks for saving CD from [Butler's] "malignant revenge"; [Ernst Ludwig] Krause's letter is very good and strictly accurate; it is lucky corr. did not come to get pigs, for infection is severe, strictest rules are observed, and as magistrate, CD must give orders daily for animals to cross roads, even if only from one field to another on same farm.[1]

1. See: Romanes, "Unconscious Memory," *Nature*, *23* (1880-1881): 285-87; and Krause, "Unconscious Memory--Mr. Samuel Butler," *Nature*, *23* (1880-1881): 288.

On pigs, see preceding letter, above.

582 1881 Feb. 11th

To /George John/ ROMANES; Down (type 9)
ALS; 8 X 5; 2p.

Thanks for second letter in Nature; is particularly pleased by first part; sympathy expressed privately and publicly toward CD almost makes CD glad that /Samuel/ Butler attacked him.[1]

1. See Romanes, "Mr. Butler's 'Unconscious Memory'," Nature, Lond., 23 (1880-1881): 335-36. On Butler controversy in general, see Barlow, ed., Autobiography, 167-219. See also: preceding letter, above; and Life and Letters, III, 220.

583 1881 Feb. 11th /pmk. FE11/ 81/

To T/homas/ R/oscoe/ R/ede/ STEBBING; Down (type 9)
ALS; 8 X 5; 4p. and env., add. / The Revd. T. R. R. Stebbing/ Tunbridge Wells/

Thanks for letter in Nature; case is well put, and corr. interprets CD's conduct accurately; has no idea how he merited such vehemence from /Samuel/ Butler; /Ernst Ludwig/ Krause's failure to mention the correction and enlargement of his text was accidental; CD failed to mention these extra efforts by Krause because CD was busy with experiments; the historical sketch appeared for the first time in third edition of /Darwin, Origin (1861)/; there was no time to put it in second edition /1860/.[1]

1. See Stebbing, "Mr. Butler's 'Unconscious Memory'," Nature, Lond., 23 (1880-1881): 336. See also preceding letter, above.

584 /1881/ March 7th

To /George John/ ROMANES; Down (type9)
ALS; 8 X 5; 4p.

Printed in full, with minor changes: More Letters, II, 213-14 (letter 548).

585 /1881/[1] March 9th

To /George John/ ROMANES; Down (type 8)
ALS; 8 X 5; 3p.

Thanks for letter;[1] corr.'s remarks on consciousness and self-consciousness are good; may have worked for weeks to no avail; CD's experiments showed that worms do learn, but CD dare not trust results; will try to eliminate source of doubt.[1]

1. This letter is almost certainly an answer to Romanes's reply to the preceding letter, above. See also More Letters, II, 215 (letter 549).

586 1881 Mar 26.

To [George John] ROMANES; Down (type 8) (last two lines crossed out)
LS; 8 X 5; 4p.

Printed in full, with minor changes: Life of Romanes, 108-09.

587 1881 Ap. 16

To [George John] ROMANES; Down (type 9)
ALS; 8 X 5; 10p.

Printed in full, with minor changes: Life of Romanes, 109-12. p. 111, line 11, change "prick" to "pith". p. 111, line 22, change "softer" to "softest".
Printed, with minor changes and minor omissions: Life and Letters, III, 243-45.

588 1881 Ap. 18th

To [George John] ROMANES; Down (type 9)
ALS; 8 X 5; 8p.

Printed, with minor changes: Life of Romanes, 114-16. p. 115, line 1 following signature, add: "I am extremely glad that you seem to have silenced [Samuel] Butler and his reviewers. But Mr Butler will turn up again, if I know the man."
Printed: More Letters, II, 431-32 (letter 431). See full text for changes.
Printed: Life and Letters, III, 206. See full text for changes.

589 1881 April 25th

To [George John] ROMANES; Down (type 9)
ALS; 8 X 5; 4p.

Printed, with minor changes: Life and Letters, III, 208-09. p. 208, line 15, change "with Miss Cobbe....." to "with the sweet Miss [Frances Power] Cobbe.-- Good Heavens what a liar she is: did you notice how in her second letter she altered what she quoted from her first letter, trusting to no one comparing the two?"[1]

1. See: Cobbe, "Mr. Darwin and Vivisection," London Times, April 19, 1881, p. 8; and idem, "Mr. Darwin on Vivisection," London Times, April 23, 1881, p. 8.

590 [1881][1] May 20th

To W[illiam Erasmus DARWIN]; Down.
ALS; 8 X 5; 2p.

Does not know what to do with £3,000 from United States funded loan; has too much in railways; sees that United States is offered $ 1 million in bonds at three and a half per cent; does this indicate that U.S. is a good security; please advise CD about what to do with the investment.

1. This letter is written in the shaky hand characteristic of CD from late 1880 to his death in April, 1882. This leaves May of 1881 as the only May in which "shaky hand" letters were written.

591 1881 Jun 17th.

To ?; Glenrhydding House/ Patterdale, Penrith[1]
ALS; 8 X 5; 1p. (enclosure wanting)

Thanks for enclosed paper; case is curious and "an awful look-out for palaeontologists"; thinks there are analogous cases with the Pulmoniferae; is surprised that author says it is not established that one mollusc can be parasitic within shell of another.

1. Down address variant, type 8, is crossed out.

592 1881 June 22

To ?; Glenridding House/ Patterdale/ Penrith[1]
LS; 8 X 5; 3p.

Suggests modification in corr.'s trials in formation of diamonds; use flesh of animal for carbon, as this may be nature's source of carbon for diamonds; has long wanted to see what chemicals resulted from slow cooling, after much heating, of all the elements one might find in living organisms; imagines a protein might be generated; this is example of "'a fool's experiment'", the likes of which CD tries now and then for fun, with interesting side-results occasionally.

1. Down address variant, type 8, is crossed out.

593 1881 June 25th [pmk. AMBLESIDE 25 JU/ 81]

To R[euben] A. BLAIR; Down (type 9)
ALS; 8 X 5; 1p. and env., add. [R. A. Blair Esqr--/ Sedalia/ Mo./ U. States]

Thanks for kind expressions; congratulations about mastodon remains; "I hope that the study of natural history may give your daughter a large share of the satisfaction which the study has given me.--"

594 [1881] July 4th

To ⌈George John⌋ ROMANES; Down (type 8)
ALS; 8 X 5; 3p.

Leaves for Penrith in a few hours; will be at Down tomorrow night; has heard from ⌈John⌋ Collier, and wife ⌈Emma Wedgwood Darwin⌋ has arranged everything ⌈for portrait of CD by Collier⌋; rejoices about corr.'s book; wants to know what corr. thinks about "the German Physiologico-Evolution book[1] & about the Creed of Science"; has read the latter with much interest; "I rejoice for my own private eating that you have taken to review & write so much in Nature, & if it does not waste too much of your time you thusly do a public service.-- Nature seems to me an excellent Journal, & I look forward weekly with pleasure to reading or skimming the whole."

1. Perhaps Wilhelm Roux, Der Kampf der Theile im Organismus... (Leipzig: 1881).

595 1881 August 5th

To ?; Down (type 9)
ALS; 8 X 5; 1p. (enclosure wanting)

Never lends money; but since corr. seems in need, encloses check for £30 as a present; this might support corr. while he is ⌈learning? leaving?⌋ the special department; cannot assist further; acknowledge.

596 ⌈1881⌋ Augt. 7th

To ⌈George John⌋ ROMANES; Down (type 9)
ALS; 8 X 5; 2p.

Printed in full, with minor changes: Life of Romanes, 120.
One sentence printed: Life and Letters, III, 223.

597 1881 Sept. 2d.

To ⌈George John⌋ ROMANES; Down (type 9)
ALS; 8 X 5; 8p.

Printed, with minor changes and minor omissions: Life of Romanes, 123-25.
At end of letter is: thanks for invitation to sons, but none of them will be in Scotland.
Printed, with changes: Life and Letters, III, 209-10.[1]

1. For letter to CD for which this letter is reply, see Life of Romanes, 121-23.

598 ⌈1881⌋ Oct 14th

To [George John] ROMANES; Down (type 9) (black border)[1]
ALS; 7 X $4\frac{1}{2}$; 3p.

Printed in full, with minor changes: Life of Romanes, 126-27.

1. CD's brother Erasmus Alvey Darwin died on August 26; see "Darwin's Journal," 21 and 21n.

599 1881 November 2d.

To ?; Down (type 9)
ALS; 8 X 5; 3p.

Thanks for informing CD of error; perhaps amount per square yard was one pound, fourteen and a half ounces instead of one pound thirteen and a half ounces; calculations were checked; two other errors have been discovered, but they are so flagrant that no one was fooled; will try to find source of error.

600 1881 Nov. 6th

To ?; Down (type 9)
ALS; 8 X 5; 3p.

Can now explain blunder about amount of earth ejected per acre; in one calculation, assumed that worms worked 45 days, while in another, assumed they worked 40 days; concluded that shorter period was correct, but forgot to go back and correct; a son of CD finds relation between weight for square yard and for acre correct as given.[1]

1. See preceding letter, above.

601 1881 Nov. 12th

To [George John] ROMANES; Down (type 9)
ALS; 8 X 5; 4p.

Printed in full, with minor changes: More Letters, I, 395-96 (letter 308).

602 [1881 November] [end. Recd. Nov 1881]

To ?; no location
AN, S by init.; $4\frac{1}{2}$ X $7\frac{1}{4}$; 2p., end. [Recd. Nov. 1881]

What does [Wilhelm Friedrich Philipp] Pfeffer mean in relation to [Julius Ritter von] Wiesner's book [Das Bewegungs-vermögen der Pflanzen... (Wien: 1881)]; Wiesner thinks process of light bending plant towards it is a direct effect, but CD thinks light is merely a stimulus that tells plant which way to bend; thought Pfeffer was on Wiesner's side;

thinks this might help corr. and Mr [Patroll?] in understanding Pfeffer's difficult style.

603 1881 Dec. 8th

To [George John] ROMANES; Down (type 9) (black border)[1]
ALS; 7 X 4½; 3p. (enclosures wanting)

Returns letters and acknowledges check for £12·10; circular is grand and [Charles Grant Blairfindie] Allen is noble; nothing can be done by subscribers; suggests giving microscope to Allen; will be in London next week and will see corr.; is overwhelmed with "stupid letters"; will send [Darwin], *Origin* [(1880)].[2]

1. CD's brother Erasmus Alvey Darwin died on August 26.
2. See Edward Clodd, *Grant Allen: A Biography* (London: Grant Richards, 1900), 84-85. See also letter to Romanes, February 3, 1880, above.

604 [1881] Dec. 10th

To [George John] ROMANES; Down (type 9)
ALS; 8 X 5; 2p.

Cannot help [James Cossar] Ewart; has already written testimonial for [Edwin] R[ay] Lankester, and has on this ground declined to give one for [William Carmichael] Mc. Intosh [*sic*; M'Intosh], "of whose work I have a very high opinion; thinks Ewart is fit; remembers interesting interview with Ewart and bacteria at University College laboratory; will be in London after Thursday [December 15]; will see corr. then.

605 1881 Dec. 16th

To ?; Down (type 9)
ALS; 8 X 5; 3p.

Letter from corr. to CD's *late* brother Erasmus [Alvey Darwin] has been forwarded to CD; Erasmus had nothing to do with the Congress; CD was not a member either, although CD went to London to be introduced to some members; was asked to Lady [Sophia Coutts Burdett, styled] Burdett-Coutts' party, but did not attend; it would therefore be inappropriate to include CD's likeness into corr.'s composition; if corr. still wishes to do so, will send "an excellent, unpublished photograph of myself, made by my son Lieut: [Leonard] Darwin"; is too busy and weak to give a sitting.

606 [1881 December][1] 17th Saturday

To [George John] ROMANES; 4 Bryanston St. (black border)[2]
ALS; 7 X 4½; 3p.

Wants to see corr. before leaving London; will see [Thomas Henry] Huxley tomorrow morning; will see [Andrew] Clark this evening; call upon CD later tomorrow or early Monday; if well, will see [John Scott] Burdon-Sanderson on Monday; "My visit to London has been rather a failure, for I have been able to do hardly anything."

1. There was only one Saturday the 17th on which CD was at Bryanston Street; see "Darwin's Journal," 19-21.
2. CD's brother Erasmus Alvey Darwin died on August 26.

607 [1881] Dec. 19th

To [F. W.] SURMAN; Down (type 9)
ALS; 8 X 5; 1p.

Please write to CD at Down, giving full name, birth date, age, duration of corr.'s having lived with CD's brother [Erasmus Alvey Darwin], and duration of corr.'s having lived at previous place, with name and address of family.

608 1881 Dec. 22d.

To [F. W.] SURMAN; Down (type 9)
ALS; 8 X 5; 2p.

Corr.'s case is hopeless, since [Albert Carl Ludwig Gotthilf] Günther says *that the man must not be beyond 30 years age*; moreover, permission of one of the three *chief* trustees must be obtained before application; has no influence on archbishop, Chancellor, and Speaker; sorry for disappointment; Günther says number of applicants is great.[1]

1. See preceding letter, above.

609 [1882] 1881 (*sic*) Jan 1st[1]

To [George John] ROMANES; Down (type 9)
ALS; 8 X 5; 8p.

Printed, with minor changes: *More Letters*, I, 389-90 (letter 302). At end of letter is: received December number of *Nineteenth Century* with corr.'s article, but has not yet read it;[2] has had note from [Charles] Grant [Blairfindie] Allen, which calls CD's attention to "capital fact about Sexual Selection" in [Nils Adolf Erik Nordenskiöld,] *Voyage of the Vega [Round Asia and Europe...*, tr. Alexander Leslie, 2v. (London: Macmillan and Co., 1881)], II, 97.

1. Mention of Brazilian Legation in this letter couples it with the two subsequent letters, below. See also date of publication in note 2, below. Apparently CD erred, as one is wont to do on the first day of

the new year, by writing down the previous year.
2. Romanes, "Intelligence of Ants II," Nineteenth Century, 10 (July-December 1881): 245-58.

610 [1882][1] Jan 3d.

To [George John] ROMANES; Down Beckenham
ALS; 4½ X 7; 1p.

Brazilian Legation says documents were intended for CD; secretary [of the Legation] is sure the Baron [de Villa Franca] would be gratified if statements were published; what does corr. think about whole story; return documents.

1. See subsequent letter, below.

611 1882 Jan 6th

To [George John] ROMANES; Down (type 9)
ALS; 8 X 5; 3p. (enclosures wanting)

Is glad to shirk extra labor, so accepts corr.'s offer to prepare paper for Linnean Society and abstract for Nature; will send copies to the Baron [de Villa Franca] and Dr. Glass [Superintendent of the Botanic Garden, Rio de Janeiro]; suggests an introduction for paper; has scribbled an outline for paper; agrees with corr. about the microscope and [Charles] Grant [Blairfindie] Allen.[1]

1. See preceding two letters. See also Baron de Villa Franca and Dr. Glass, "On New Varieties of the Sugar-cane Produced by Planting in Apposition," Proc. Linn. Soc. Lond., November 1880-June 1882, 30-31. On Allen and the microscope, see letter to Romanes, December 8, 1881, above.

612 1882. Jan. 20th

To [George John] ROMANES; Down (type 9)
ALS; 8 X 5; 2p.

Sees propriety of [George] Croom Robertson's proposal; agrees to all that corr. proposes; cannot personally see the advantage of waiting until another half-payment is made, prefers to see testimonial or present given at once; does not wish to impose CD's judgment on Robertson.[1]

1. This letter apparently refers to the giving of a microscope to Charles Grant Blairfindie Allen; see preceding letter, above.

613 [1882 (?January)][1] 23d

To /George John/ ROMANES; Down.
ANS; 4½ X 7; 1p.

Is glad about microscope; H/erbert/ Spencer's plan would have destroyed graciosity of the present; is better, but is still weak; hopes corr. will be able to read sugar cane paper soon.

1. This letter obviously follows shortly after the preceding letter. For sugar cane paper, see letter to Romanes, January 6, 1882, above.

614 1882 Jan. 28th

To /George John/ ROMANES; Down (type 9)
ALS; 8 X 5; 3p.

Is bound to do what /Donald/ McAlister wishes, even though it is superfluous; can only manage to write one or two poorly-done pages; has just received corr.'s note of 25th; declines again to provide testimonial to /James Cossar/ Ewart, since CD had already given one for /Edwin Ray/ Lankester and had refused to give one to /William Carmichael/ McIntosh /i.e. M'Intosh/; P/atrick/ Geddes has also done excellent work, but CD supposes he is very young; according to /Alexander/ Dickson's testimonial to Geddes, /John/ Young is also in the field; sorry to disappoint corr.[1]

1. See letter to Romanes, December 10, 1881, above.

615 1882 Feb 4th.

To ?; Down (type 9)
ALS; 8 X 5; 3p., sketches (enclosure wanting)

Make copies on wood of the cuts enclosed in blue lines in plate to the enclosed paper;[1] gives directions for execution of copies; "They must be fac-similes"; needs them as soon as possible for paper to be read before Linnean Society;[2] sketch shows what copies should look like when done; return the pamphlet; will soon have another sketch engraved.

1. Francis Darwin, "The Process of Aggregation in the Tentacles of Drosera rotundifolia," Q. Jl. microsc. Sci., 16 (1876): 309-19.
2. Darwin, "The Action of Carbonate of Ammonia on Chlorophyll-bodies," J. Linn. Soc. (Botany), 19 (1882): 262-84.

616 /1882/ Feb. 9th /end. 11 FEB 1882/

To /Trübner & Co./; Down (type 7)
AL in third person; 4 3/4 X 7½; 1p., end. by stamp /Trübner & Co./ 11 FEB 1882/ London/

Orders International Sc/‾ientific_7 Dict/‾ionary_7 to be sent by post.

617 1882. March 29th

To ?; Down, Kent
AN in third person S; 8 X 5; 1p.

CD thinks Mr /‾Loeb? Loct?_7 ought to have sent an addressed envelope.

618 1882 April 3d

To /‾Philip Lutley_7 SCLATER; Down (type 9)
ALS; 8 X 5; 1p. and end. /‾733_7 (enclosure wanting)[1]

Sends enclosed paper to be read at Zoological Society and to be printed in the Journal, if referees recommend; if paper is rejected, would appreciate its return, so it can be submitted to Nature; wants its contents recorded.[1]

1. Paper is probably W. Van Dyck, "On the Modification of a Race of Syrian Street-Dogs by Means of Sexual Selection, with a Preliminary Notice by Charles Darwin," Proc. zool Soc. Lond., 1882, 367-70.

619 1882 April. 17th

To ?; Down (type 9)
ALS; 8 X 5; 3p.

Corr. has misunderstood CD, but mistake was a natural one and the criticism was good; ought not to have interpolated the sentence about the burying of food;[1] case was mentioned to illustrate a long-continued habit, since CD thinks well-fed domestic dogs do not revisit their buried treasures; when burying food, dog makes hole with front legs alone and shovels in the earth with his nose; there is no resemblance to excrement-covering movements; thanks for courteous expressions.

1. See Darwin, Variation under Domestication (1868), I, 182.

1882 April 19--CD died at four in the afternoon.

620 n.y. Feb. 21

To "My dear Camilla"; Down.
ALS; $4\frac{1}{2}$ X 7; 1p. (enclosure wanting)

Look at enclosed letter from an "odd & good-hearted man"; no one at Down can make it out; do not translate whole letter, but only part in which the man recommends some treatment; doubts if CD should try it; does writer of the letter seem to know what is the matter with CD?

621 n.y. 14th

To [John William] LUBBOCK, [Baron Avebury]; Down.
ALS; 7 3/4 X 4 3/4; 3p.

Received yesterday enough specimens of *Helix pomatia* [edible snail] "for my Snailery"; thanks for Lepidoptera books, which CD's sons study with interest; give number of acres in [Larch?] wood, and when planted; have collected in it about 115 species of plants; Leonard Horner comes to Down on Monday; if corr. would like to meet Horner, come for supper on Tuesday; the Carters (staying at Colonel [Caters?] house) and [John Brodie] Innes and wife.

622 n.d.

To G. DENNY; no location
Address leaf only, which reads "G. Denny Esqr/ Philosophical Hall/ Leeds"; 2 X 7

623 n.d.

To [John PHILLIPS]; no location
ALS; 6½ X 5; 1p., incomplete

Has not yet received book, but thanks for note; book will surely interest CD.

624 n.d.

To G[eorge] J[ohn] ROMANES; Down, Beckenham.
AN, S by init., on p.c.; 1p. and add. [G. J. Romanes Esqr/ Dunskaith/ Nigg/ Rosshire]

Has not read corr.'s letter; asks permission to show it to George [Howard Darwin], who arrived recently, and to CD's other sons; will tell them to keep it secret if corr. wishes, but a secret cannot forever be well kept owing to forgetfulness.

625 n.d.

To [George John ROMANES]; no location
?AN; 8 X 5; 1p.

"18 Cornwall Terrace [Romanes's address]/ Linn. Soc. & List/ Carrots--Rabbits--gray same side/ Copy of Var. under Dom./ fuller [abstract?] of Potato-grafts/ [Francis] Galton's article on Pangenes./ Medusa & Royal Soc/ 11. York Place"[1]

1. See: Darwin, Variation under Domestication (1875), I, 420-22; and Galton, "Experiments in Pangenesis, by Breeding from Rabbits of a Pure Variety, into Whose Circulation Blood Taken from Other Varieties Had Previously Been Largely Transfused," Proc. R. Soc., 19 (1871): 393-410.

626 n.d.

To ?; no location
A fragment of L; 1 3/4 X 6; 2p.

"I am greedy for facts.--"

## APPENDIX

## OTHER DARWIN MATERIALS IN THE LIBRARY OF THE AMERICAN PHILOSOPHICAL SOCIETY

In addition to manuscript letters to and from Charles Darwin, several other items in the Library of the American Philosophical Society may be of interest to the Darwin scholar. These are discussed below in more-or-less general way which, although not a comprehensive list, should give readers some idea of the nature of these collections. The items are discussed in groups, by type of document.

### Manuscript items

Included under this rubric are three types of documents: 1) the correspondence and other papers of those people who corresponded with Darwin--all of it relating directly to manuscript letters calendared above; 2) autograph documents from Darwin's own papers or from others, but relating directly to Darwin and not connected to any of the manuscript letters calendared above; and 3) manuscript correspondence to and from other members of the Darwin family besides Charles Darwin.

There are seven items of the first type. All of these pertain to the case of the inheritance of an injury in a goose, as related to Darwin by Reuben A. Blair in the correspondence with him calendared above. The documents are: 1) a photograph of the deformed goose; 2) a letter from Blair concerning the goose and printed in the Sedalia _Democrat_; 3) a letter from William Henry Flower to Blair; 4) the report by Flower and his assistant, Dr. Larson, on the wings of the affected geese; 5) a letter from Blair to an unnamed correspondent; and 6) letters exchanged between Blair and Spencer Fullerton Baird of the Smithsonian Institution.

There are twenty items of the second type. These documents are: 1) five leaves from the manuscript of the _Origin of Species_; 2) a signed page from the manuscript of the _Descent of Man_; 3) a signed page from some other Darwin manuscript (discusses _Catasetum_); 4) a printed petition for the endowment of research, issued by J. Norman Lockyer and C. E. Appleton, and signed by Darwin; 5) the final page of the petition for a pension for Alfred Russel Wallace, signed by twelve persons, including Darwin; 6) sketches of Darwin, his wife, grandson, and dog, by Albert Goodwin; 7) documents pertaining to Darwin's funeral in Westminster Abbey (eleven items); 8) inscription from the statue of Darwin at Oxford University; 9) a discussion of Darwin's religion by H. Buxton Forman; and 10) a pass to the Zoological Gardens, issued to James Gough by Darwin.

There are 44 items of the third type, which is too many to list individually here. Members of the Darwin family involved in this correspondence are: Emma Wedgwood Darwin (two letters, correspondence with Mrs. Georgiana Rosetta Smyth Flower and with John Maurice Herbert); Francis Darwin (39 letters, correspondence with Leo Abram Errera, George John Romanes, Thomas Roscoe Rede Stebbing, and Otto Zacharias); and George Howard Darwin (three letters, correspondence with F. W. Surman and Otto Zacharias).

## Photocopies of manuscript items

The Library has 26 documents pertaining to Darwin, the originals of which are located elsewhere; only photographically-reproduced paper copies of the documents are at the APS. These fall into two categories: photocopies of manuscript correspondence with Darwin; and photocopies of Darwin-related documents from the papers of Darwin correspondents. In the first category, there are copies of correspondence with the College of Physicians of Philadelphia (one letter), Ignatius Donnelly (two letters), Hugo de Vries (one letter ), Benjamin Dann Walsh (eighteen letters), and Jeffries Wyman (two letters). In the second category, there is a photocopy of a letter from Frank J. Mead, editor of the Minneapolis Evening Times, to Ignatius Donnelly, concerning Darwinism and Christianity, and a photocopy of a portion of Donnelly's diary which mentions Darwin.

## Microfilms of manuscript items

The Library has nine sets of microfilms of Darwin materials which are on deposit at various institutions all over the world. These films contain the following: 1) correspondence in possession of Down House, Kent (one reel); 2) correspondence in the Robert Stecher Collection at the Cleveland Medical Library Association (one reel); 3) correspondence in the New York Botanical Garden (one reel); letters to Auguste H. Forel, from originals in the Medicinhistorisches Institut, University of Switzerland, and in University of Basel Library (sixteen frames); 4) letters to Bernhard Studer and to A. von Morlot (seven items); 5) letters to J. Moulinie, A. Dohrn, Karl Christoph Vogt, Pictet, and de la Rive, from originals in the Bibliotheque Publique et Universitaire de Geneve (25 letters); 6) notebooks written during the voyage of the Beagle, from Down House, Kent (eighteen notebooks, one reel); 7) diary and correspondence written during the voyage of the Beagle, from Down House, Kent (one reel); and 9) chronology and bibliography of Darwin's life, compiled by Sir Gavin de Beer, containing an incomplete list of all of Darwin's correspondence (two reels).

## Photographs and prints

The Library has approximately twelve different photographs and prints of Darwin and of Down House. Some of the best of these are used as illustrations in this calendar.

Books

There are about 150 copies of various editions of Darwin's works in the Library, plus a fair collection of some of the rarer and/or more significant editions of the works of his correspondents, colleagues, and contemporary naturalists and biologists. Space does not permit a full listing of the 150 works by Darwin in the Library, but it is possible to give some indication of the collection by referring to the number assigned to each edition held by the Library in the standard bibliography of Darwin's works (R. B. Freeman, The Works of Charles Darwin: An Annotated Bibliographical Handlist [London: Dawsons of Pall Mall, 1965]). The "Freeman numbers" of the Library's holdings are given below, without comment.

1, 2, 3, 4, 5, 7, 8, 10, 16, 83, 84, 85, 86, 87, 88, 89, 90, 91, 95, 99, 103, 105, 106, 108, 109, 110, 111, 112, 113, 114, 115, 116, 117, 142, 205, 207, 208, 210, 219, 220, 231, 231n, 232, 234, 246, 247, 250, 255, 269, 281, 283, 292, 297, 298, 303, 305, 308, 309, 311, 314, 315, 319, 332, 333, 334, 343, 344, 345, 346, 349, 359, 361, 362, 364, 365, 366, 367, 369, 370, 371, 173, 374, 375, 376, 377, 378, 380, 381, 383, 396, 405, 406, 407, 409, 411, 416, 425, 426, 432, 445, 447, 452, 454, 458, 461, 464, 468, 471, 473, 476, 477, 492, 503, 507, 508, 512, 513, 517, 530, 541.

In addition to these editions, the Library has at least 26 editions of Darwin's works not listed in Freeman.

Subject card catalogs

For a great many years, the Library has maintained subject card catalogs on Darwinism. These catalogs list any published work among the Library's holdings which mentions or discusses Darwin and/or Darwinism. They provide an unparalleled source for study of the influence of Darwin upon society and upon Western thought. There is one card catalog for books, containing approximately 2,800 entries, and another for journal articles, containing approximately 1,000 entries.

An effort has been made to make this appendix as complete as possible, but as the Library is constantly adding to its Darwin holdings, the careful scholar is warned that this compendium will be out of date in a short time.

## APPENDIX

## OTHER DARWIN LETTERS IN THE GREATER PHILADELPHIA AREA

It is expected that this calendar will render unnecessary many visits to Philadelphia by Darwin scholars which would have been mandatory otherwise. While this is a beneficial result of the publication of this book, it is not without its harmful side effects. The worst of these would have been that the many miscellaneous Darwin materials in other institutions in the Philadelphia area might be neglected; this appendix is designed to prevent this by listing the results of a search by mail for other manuscript Darwin letters in select institutions in the greater Philadelphia area.

To conduct the search, a form letter was sent to the 31 institutions in the area judged by the editor to be the most likely to possess Darwin letters. Thirty institutions replied; they are listed below, and the name of the person responding is given for each institution.

Academy of Natural Sciences of Philadelphia (Martha T. Pilling, Library Assistant); Bryn Mawr College Library (James Tanis, Head Librarian); Bucknell University (George M. Jenks, Librarian, Ellen Clarke Bertrand Library); College of Physicians of Philadelphia (Ellen G. Gartrell, Assistant Curator, Historical); Dickenson College (Danna Spitzform, Assistant to the Curator, Special Collections, Boyd Lee Spahr Library); Drexel University (Michael Halperin, Archives and Special Collection); Eleutherian Mills Historical Library (Betty-Bright P. Low, Research and Reference Librarian); Franklin and Marshall College (Frances L. Hopkins, Reference Librarian, Fackenthal Library); The Franklin Institute (Stephanie A. Morris, Associate Archivist); Free Library of Philadelphia (Howell J. Heaney, Rare Book Librarian); Hahnemann Medical College Library (Barbara Williams, Acting Librarian); Haverford College Library (Edwin B. Bronner, Librarian); Historical Society of Delaware (Gladys M. Coghlan); Jefferson Medical College Library (Robert T. Lentz, Librarian); Lafayette College (Ronald E. Robbins, Reference Librarian, David Bishop Skillman Library); Lehigh University (James D. Mack, Director of University Libraries, Linderman Library); Library Company of Philadelphia (Edwin Wolf, II, Librarian); Pennsylvania State University (Dorrie Evans, Rare Books and Special Collections, Fred Lewis Pattee Library); Philadelphia College of Pharmacy and Science Library (Elizabeth W. J. Chase, Librarian); Princeton University Library (Mardel Pacheco, Assistant to Curator of Manuscripts); Philip H. and A. S. W. Rosenbach Foundation Museum and Library (Clive E. Driver, Director); Saint Joseph's College (Josephine Savaro, Head Librarian, Drexel Library); Swarthmore College (Judith Pullam, Administrative Assistant, Thomas E. and Jeanette L. McCabe Library); Temple University (Thomas M. Whitehead, Head, Special Collections Department, Samuel Paley Library); University of Delaware Libraries (Stuart Dick, Special Collections, Hugh M. Morris Library);

University of Pennsylvania Libraries (Neda Westlake, Rare Book Collection); Ursinus College (Calvin D. Yost, Jr., Librarian, Myrin Library); Villanova University (Mary A. Dorrian, Readers Service, Falvey Memorial Library); Wagner Free Institute of Science (Robert Chambers, Director); and Widener College (Lee C. Brown, Librarian, Wolfgram Memorial Library).

The editor is grateful for the cooperation of these institutions and individuals.

Twenty-four letters of Charles Darwin were produced by the search. They are listed below, in chronological order.

1. 1838 January 23 Sunday, to John Stevens Henslow; Al, S by init.; 4p.; courtesy of the Historical Society of Pennsylvania.
2. 1843 July 19th, to Ernest Dieffenbach; ALS; 2p.; courtesy of the Historical Society of Pennsylvania.
3. 1853 Jan. 10th, to Albany Hancock; ALS; 4p.; courtesy of the Historical Society of Pennsylvania.
4. ?1854 Nov. 20th, to John Stevens Henslow; ALS; 2p.; courtesy of the Princeton University Library.
5. 1860 March. 4th, to Joseph Leidy; ALS; 4p.; courtesy of the Academy of Natural Sciences of Philadelphia (see their Collection 1). See calendar entry for this letter (number 202), page 69, above.
6. 1860 May 8., to the Secretary of the Academy of Natural Sciences of Philadelphia; ALS; 1p.; courtesy of the Academy of Natural Sciences of Philadelphia (see their Collection 330).
7. 1862 April 25th, to Heinrich Georg Bronn; ALS; 6p.; courtesy of Lehigh University.
8. 1871 July 1, to ?; ALS; 1p.; courtesy of the Historical Society of Pennsylvania.
9. 1872 Oct. 10th, to Alpheus Hyatt; ALS; 4p.; courtesy of the Princeton University Library (see the Hyatt and Mayer Correspondence).
10. 1873 Feb 19, to William M. Canby; LS; 3p.; William M. Canby Correspondence, Society of Natural History of Delaware Archives, on deposit at the Historical Society of Delaware.
11. 1873 May 7, to William M. Canby; LS; 2p.; William M. Canby Correspondence, Society of Natural History of Delaware Archives, on deposit at the Historical Society of Delaware.
12. 1874 May 11, to Thomas Lauder Brunton; LS; 4p.; courtesy of the Princeton University Library.
13. 1874 June 30., to ?; LS; 2p.; courtesy of the College of Physicians of Philadelphia.
14. 1875 Oct. 30th, to Messrs. Smith & Elder; ALS; 1p.; courtesy of Lehigh University.
15. 1876 Aug. 21st, to Messrs. Smith & Elder; ALS; 1p.; courtesy of Lehigh University.
16. 1876 Nov. 20th, to ?; LS; 2p.; original in private possession of Dr. Seymour Adelman, c/o James Tanis, Head Librarian, Bryn Mawr College Library.
17. 1877 March 7., to Messrs. Smith & Elder; LS; 4p.; courtesy of Lehigh University.

18. 1877 June 6th, to ?; ALS; 1p.; courtesy of Haverford College Library (Charles Roberts Autograph Letters Collection).
19. 1880 Nov. 5, to ?; ALS; 1p.; courtesy of the Historical Society of Pennsylvania.
20. 1882 Feb 22, to ?; LS; 4p.; courtesy of Princeton University Library (privately owned; from the private library of William H. Scheide, Princeton, New Jersey; no connection with Princeton University Library).
21. n.y. April 6th., to ?; ALS; 3p.; courtesy of Lehigh University.
22. n.y. May 7, to ?; ALS; 1p.; courtesy of Philip Schwartz, M.D., Medical Research Director, Warren State Hospital, Warren, Pennsylvania (privately owned by Dr. Schwartz).
23. n.y. Oct 2d., to ?; ALS; 1p.; courtesy of Lehigh University.
24. n.y. Oct 14th, to "Madam"; ALS; 2p.; courtesy of Lehigh University.

All letters are listed with the permission of their owners, for which permission the editor is grateful.

## BIOGRAPHICAL NOTES AND NAME INDEX

Numbers in the index below refer to letter numbers in the left margins, not to page numbers. In the case of each person listed, some effort was made to provide the following: full name; dates of birth and death; nationality (if not a British subject); occupation(s), using somewhat anachronistic terminology at times; specific biographical information of importance in establishing a link between the person and Charles Darwin; and date elected as Fellow of the Royal Society of London or other major scientific body, if any. Letters to each person, letters from each person, and letters in which each person is mentioned are then listed. Only persons from Charles Darwin's time or earlier are indexed. Whitfield J. Bell, Jr., Marcella P. Carroll, Nancy E. Carroll, and David Levine contributed substantially to the creation of this index; they have the editor's sincere thanks.